D1233986

Sustainability
in the Food Industry

The *IFT Press* series reflect the mission of the Institute of Food Technologists — to advance the science of food contributing to healthier people everywhere. Developed in partnership with Wiley-Blackwell, *IFT Press* books serve as leading-edge handbooks for industrial application and reference and as essential texts for academic programs. Crafted through rigorous peer review and meticulous research, *IFT Press* publications represent the latest, most signifi cant resources available to food scientists and related agriculture professionals worldwide.

Founded in 1939, the Institute of Food Technologists is a nonprofi scientifi society with 22,000 individual members working in food science, food technology, and related professions in industry, academia, and government. IFT serves as a conduit for multidisciplinary science thought leadership, championing the use of sound science across the food value chain through knowledge sharing, education, and advocacy.

A John Wiley & Sons, Ltd., Publication

Sustainability in the Food Industry

EDITOR

Cheryl Baldwin

A John Wiley & Sons, Ltd., Publication

Cheryl J. Baldwin, PhD, is Vice President of Science and Standards for Green Seal, a Washington D.C.-based non-profi organization dedicated to safeguarding the environment by transforming the marketplace by promoting the manufacture, purchase, and use of environmental standards and providing independent certificatio to those standards. Previously, Dr Baldwin served as Program Leader and Senior Research Scientist at Kraft Foods, Glenview, IL.

Edition firs published 2009
©2009 Wiley-Blackwell and the Institute of Food Technologists

Blackwell Publishing was acquired by John Wiley & Sons in February 2007. Blackwell's publishing program has been merged with Wiley's global Scientific Technical, and Medical business to form Wiley-Blackwell.

Editorial Office
2121 State Avenue, Ames, Iowa 50014-8300, USA

For details of our global editorial offices for customer services, and for information about how to apply for permission to reuse the copyright material in this book, please see our website at www.wiley.com/wiley-blackwell.

Library of Congress Cataloguing-in-Publication Data
Sustainability in the food industry / [edited by] Cheryl Baldwin. – 1st ed.
   p. cm.
  Includes bibliographical references and index.
  ISBN 978-0-8138-0846-8 (hardback : alk. paper)
  1. Food industry and trade. 2. Food supply. 3. Sustainable
agriculture. I. Baldwin, Cheryl.
  HD9000.5.S828 2009
  338.1–dc22

                                      2008040237

A catalogue record for this book is available from the U.S. Library of Congress.

Set in 11.5/13.5 pt Times New Roman by Aptara® Inc., New Delhi, India

Titles in the *IFT Press* series

- *Accelerating New Food Product Design and Development* (Jacqueline H. Beckley, Elizabeth J. Topp, M. Michele Foley, J.C. Huang, and Witoon Prinyawiwatkul)
- *Advances in Dairy Ingredients* (Geoffrey W. Smithers and Mary Ann Augustin)
- *Biofilms in the Food Environment* (Hans P. Blaschek, Hua H. Wang, and Meredith E. Agle)
- *Calorimetry and Food Process Design* (Gönül Kaletunç)
- *Food Ingredients for the Global Market* (Yao-Wen Huang and Claire L. Kruger)
- *Food Irradiation Research and Technology* (Christopher H. Sommers and Xuetong Fan)
- *Food Laws, Regulations and Labeling* (Joseph D. Eifert)
- *Food Risk and Crisis Communication* (Anthony O. Flood and Christine M. Bruhn)
- *Foodborne Pathogens in the Food Processing Environment: Sources, Detection and Control* (Sadhana Ravishankar and Vijay K. Juneja)
- *Functional Proteins and Peptides* (Yoshinori Mine, Richard K. Owusu-Apenten, and Bo Jiang)
- *High Pressure Processing of Foods* (Christopher J. Doona and Florence E. Feeherry)
- *Hydrocolloids in Food Processing* (Thomas R. Laaman)
- *Microbial Safety of Fresh Produce: Challenges, Perspectives and Strategies* (Xuetong Fan, Brendan A. Niemira, Christopher J. Doona, Florence E. Feeherry, and Robert B. Gravani)
- *Microbiology and Technology of Fermented Foods* (Robert W. Hutkins)
- *Multivariate and Probabilistic Analyses of Sensory Science Problems* (Jean-François Meullenet, Rui Xiong, and Christopher J. Findlay)
- *Nondestructive Testing of Food Quality* (Joseph Irudayaraj and Christoph Reh)
- *Nanoscience and Nanotechnology in Food Systems* (Hongda Chen)
- *Nonthermal Processing Technologies for Food* (Howard Q. Zhang, Gustavo V. Barbosa-Cànovas, V.M. Balasubramaniam, Editors; C. Patrick Dunne, Daniel F. Farkas, James T.C. Yuan, Associate Editors)
- *Nutraceuticals, Glycemic Health and Type 2 Diabetes* (Vijai K. Pasupuleti and James W. Anderson)
- *Packaging for Nonthermal Processing of Food* (J.H. Han)
- *Preharvest and Postharvest Food Safety: Contemporary Issues and Future Directions* (Ross C. Beier, Suresh D. Pillai, and Timothy D. Phillips, Editors; Richard L. Ziprin, Associate Editor)
- *Processing and Nutrition of Fats and Oils* (Ernesto M. Hernandez and Afaf Kamal-Eldin)
- *Processing Organic Foods for the Global Market* (Gwendolyn V. Wyard, Anne Plotto, Jessica Walden, and Kathryn Schuett)
- *Regulation of Functional Foods and Nutraceuticals: A Global Perspective* (Clare M. Hasler)
- *Sensory and Consumer Research in Food Product Design and Development* (Howard R. Moskowitz, Jacqueline H. Beckley, and Anna V.A. Resurreccion)
- *Sustainability in the Food Industry* (Cheryl J. Baldwin)
- *Water Activity in Foods: Fundamentals and Applications* (Gustavo V. Barbosa-Cànovas, Anthony J. Fontana, Jr, Shelly J. Schmidt, and Theodore P. Labuza)
- *Whey Processing, Functionality and Health Benefits* (Charles I. Onwulata and Peter J. Huth)

Contents

Contributors

Cheryl Baldwin (Introduction, Chapters 7, 9, and 11)
Green Seal, Inc., Washington, DC 20036

Aaron L. Brody (Chapter 4)
Packaging/Brody Inc. Duluth, GA 30095

Jeff Chahley (Chapter 8)
Kraft Foods, Northfiel , IL 60093

Ben Champion (Chapter 3)
Department of Geography, Kansas State University, Manhattan, KS 66506

Cheri Chastain (Chapter 8)
Sustainability Coordinator, Sierra Nevada Brewing Co. Chico, CA 95928

Tim Crosby (Chapter 3)
Growing Washington, Edmonds, WA 98020

Randi Dalgaard (Chapter 5)
Department of Agroecology and Environment, Faculty of Agricultural Sciences, University of Aarhus, Tjele, Denmark

Tran Thi My Dieu (Chapter 2)
Department of Environmental Technology and Management, Van Lang University, Ho Chi Minh City, Vietnam

Charles Francis (Chapters 1 and 6)
Department of Agronomy and Horticulture, University of Nebraska, Lincoln, NE 68583

Holly Givens (Chapter 7)
Organic Trade Association, Washington, DC 20008

Niels Halberg (Chapter 5)
Danish Research Centre for Organic Food and Farming DARCOF, Tjele, Denmark

Barbara Haumann (Chapter 7)
Organic Trade Association, Washington, DC 20008

John E. Hermansen (Chapter 5)
Department of Agroecology and Environment, Faculty of Agricultural Sciences, University of Aarhus, Tjele, Denmark

Sara Kaplan (Chapter 3)
Leopold Center for Sustainable Agriculture, Iowa State University, Ames, IA 50011

Lisbeth Mogensen (Chapter 5)
Department of Agroecology and Environment, Faculty of Agricultural Sciences, University of Aarhus, Tjele, Denmark

Rich Pirog (Chapter 3)
Leopold Center for Sustainable Agriculture, Iowa State University, Ames, IA 50011

Rebecca Rasmussen (Chapter 3)
Leopold Center for Sustainable Agriculture, Iowa State University, Ames, IA 50011

Amarjit Sahota (Chapter 7)
Organic Monitor, London, United Kingdom

Kantha Shelke (Chapter 6)
Corvus Blue, Chicago, IL 60610

B. Gail Smith (Chapters 5 and 8)
Sustainable Agriculture Scientist, Unilever Research, United Kingdom

John Turenne (Chapter 10)
Sustainable Food Systems, Wallingford, CT 06492

J.C. Vis (Chapters 5 and 8)
Sustainable Agriculture Programme Director, Unilever N.V.,
the Netherlands

Justin Van Wart (Chapters 1 and 6)
Department of Agronomy and Horticulture, University of Nebraska,
Lincoln, NE 68583

Nana T. Wilberforce (Chapter 11)
Green Seal, Inc. Washington, DC 20036

INTRODUCTION

Cheryl Baldwin

The food supply chain affects every individual on the planet. As a result, sustainable development of the food supply chain is imperative. Sustainable development has been define as meeting "the needs of the present without compromising the ability of future generations to meet their needs" (WBCSD, 2000). The food supply chain, also called the food industry or food system, includes aspects from production of the food, processing, distribution, consumer purchase, consumer use, and end of life. A sustainable food supply would then mean that food is produced and consumed in a way that supports the well-being of generations.

The current food supply has demonstrated impacts that make it unsustainable. Such impacts include overreliance on inputs for food production such as high-intensity animal production and production of produce out of season. For example, the supply chain contributes significant y to climate change, with agricultural production alone responsible for 17–32% of global greenhouse gas emissions (Bellarby et al., 2008). It has been estimated that the food system consumes close to 16% of the total energy use in the United States (Hendrickson, 1996). Food processing also constitutes 25% of all water consumption worldwide and 50–80% of all water used in industrial countries (Okos). Further, there remains widespread malnutrition, both under- and over-nutrition. As a result, the key sustainability considerations for the food supply include energy, waste, water, air, climate, biodiversity, food quality, food quantity, food price, food safety, employment, and employee welfare (Kramer and Meeusen, 2003). These issues, along with others, are discussed in detail in the chapters of this book.

The food industry has the capability to provide safe, nutritious, and fl vorful foods to a range of consumers. Agricultural production can provide a range of commodities for nourishment. The processing of commodities can provide a means to preserve foods for appropriate distribution and storage and also may reduce total waste by preparing commodities consumption. Distribution then can enable the food to reach those who need it.

The Strategy for Sustainable Farming and Food in the United Kingdom developed the following key principles for a sustainable food chain (DEFRA, 2006):

- Produce safe, healthy products in response to market demands, and ensure that all consumers have access to nutritious food and to accurate information about food products.
- Support the viability and diversity of rural and urban economies and communities.
- Enable viable livelihoods to be made from sustainable land management, both through the market and through the payments for public benefits
- Respect and operate within the biological limits of natural resources (especially soil, water, and biodiversity).
- Achieve consistently high standards of environmental performance by reducing energy consumption, minimizing resource inputs, and using renewable energy wherever possible.
- Ensure a safe and hygienic working environment and high social welfare and training for all employees involved in the food chain.
- Achieve consistently high standards of animal health and welfare.
- Sustain the resource available for growing food and supplying other public benefit over time, except where alternative land uses are essential to meet other needs of society.

As a result, agricultural production should be focused on providing the most nutritionally dense options with the least intensity. Food processors and manufacturers need to include sustainable actions like waste reduction and recovery, composting, recycling, and processing with minimal water and energy use. Distribution should be as efficien as possible.

The benefit of sustainable practices are important for the global social and environmental benefit mentioned, but the World Business Council for Sustainable Development has also found that businesses that incorporate sustainable practices have had greater financia success

(WBCSD, 2002). Benefit of sustainable practices include lower production costs, improved product function and quality, increased market share, improved environmental performance, improved relationships with stakeholders, and lower risks.

Consumer interest in sustainable food has grown. This interest has been attributed to the desire to improve one's personal and family health and safety (Sloan, 2007). Environmental reasons have remained a secondary benefit and in many ways unknown to consumers. For example, in a survey conducted by the Leopold Center for Sustainable Agriculture in 2007, 88% of respondents perceived local and regional food systems to be somewhat safe or very safe and had purchase preferences for such food, compared to only 12% perceiving global foods as safe (Pirog and Larson, 2007). The survey also showed that the respondents, however, did not know that airplane transport of food emitted more greenhouse gases than trucks (on a per pound basis of product transported) (Pirog and Larson, 2007).

The impacts of an unsustainable food supply on health and food safety are discussed in this book, and go well beyond the average consumers' knowledge. This indicates that as consumers learn more, their interest in sustainable food will only increase. Further, it is already evident that environmental concerns are moving higher in priority to many consumers.

This book will evaluate the sustainability of each of the main supply chain components of the food industry. There will be emphasis on environmental considerations given its significanc and need for progress. Finally, the last chapter (Chapter 11) will bring the discussion from all the chapters/supply chain components together to outline sustainability principles for food and beverage products, including strategies on how to develop/innovate more sustainable products.

References

Bellarby, J., B. Foereid, A. Hastings, and P. Smith. 2008. *Cool Farming: Climate Impacts of Agriculture and Mitigation Potential*. Amsterdam, the Netherlands: Greenpeace.

DEFRA (Department for Environment, Food and Rural Affairs). 2006. *Food Industry Sustainability Strategy*. London, UK: DEFRA.

Hendrickson, J. 1996. *Energy Use in the U.S. Food System: A Summary of Existing Research and Analysis*. Madison, WI: University of Wisconsin, Center for Integrated Agricultural Systems.

Kramer, K., and M. Meeusen. 2003. "Sustainability in the agrofood sector." In: *Life cycle Assessment in the Agri-food Sector: Proceeding from the 4th International Conference, October 6–8, 2003*, Bygholm, Denmark.

Okos, M. *Developing Environmentally and Economically Sustainable Food Processing Systems*. Available from https://engineering.purdue.edu/ABE/Research/research95/okos.sohn.96.whtml. Accessed April 5, 2008.

Pirog, R., and A. Larson. 2007. *Consumer Perceptions of the Safety, Health, and Environmental Impact of Various Scales and Geographic Origin of Food Supply Chains*. Ames, IA: Leopold Center for Sustainable Agriculture. September 2007.

Sloan, E. 2007. New shades of green. *Food Technol.* 61(12):16.

WBCSD (World Business Council for Sustainable Development). 2002. *The Business Case for Sustainable Development*. Available from http://www.wbcsd.org/web/publications/business-case.pdf. Accessed September 19, 2008.

Sustainability
in the Food Industry

Chapter 1

Agriculture

Charles Francis and Justin Van Wart

Introduction: Human Food Supply Is a Continuing Challenge

Development of a sustainable agriculture and food system must be an essential part of our long-term economic and environmental planning. Adequate food and a livable environment are both critical to the long-term survival of our species.

Research and development over the past century have provided an impressive and even unexpected surge in production of food across the prime agricultural regions, especially those with potential for irrigation. Adding to these gains have been the extraordinary contributions of plant breeding to high-input production systems and the corresponding advances in fertility and pest management. The fruits of the Green Revolution and the impacts of the International Agricultural Research Centers provide evidence of what a focused public domain program can achieve. At the same time, such an acceleration in food production has come at a price. As with any biological population, human numbers have increased in response to available food and other resources. Human population is likely to reach the current projection of 9.6 billion before it is predicted to level and drop (Brown, 2008). The increasing human population and demands for food, fuel, and other products that depend on nonrenewable natural resources have put an unprecedented pressure on the global life-support system (Tilman et al., 2002). Human activities currently exploit over 40% of total net primary productivity captured by photosynthesis, leaving just over half for the maintenance of all other species.

3

Perhaps the fragility of the global ecosystem is best illustrated by the current rate of extinction of plant and animal species. Economist and author, Lester Brown, states that we are presently in the midst of the sixth major extinction event in the earth's history, the last of which occurred 65 million years ago, wiped out the dinosaurs, and was likely the result of an asteroid hitting the planet (Brown, 2008). Today's problem is the firs such event that is almost entirely a result of human activity and our destruction of habitat. One of the immediate economic and food system impacts is the disappearance of fis and collapse of the fishin industry, with 75% of commercial fis species being removed at unsustainable rates (FAO, 2007). More far-reaching consequences of these human activities include the losses of life-sustaining ecosystem-support services (Daily, 1996). We must acknowledge that our expansion in human population and increase in food production do come at a cost, often one that we are unable to calculate.

Thus, we appear to be reaching a tipping point in the balance between exploitation of natural resources and satisfaction of human wants and needs. Brown's *Plan B 3.0: Mobilizing to Save Civilization* (Brown, 2008), provides an overview of current challenges as well as potential solutions at the global scale. Also, the recent book *Developing and Extending Sustainable Agriculture: A New Social Contract* (Francis et al., 2006) provides an up-to-date catalog of sustainable practices in agriculture, and serves as another prime resource for this chapter.

Technical Research in Agriculture

The agricultural advances during the past century were truly spectacular. While human population increased from 3 to 6 billion people between 1960 and 2000, food grain production increased by a factor of three, easily keeping up with population in aggregate and solving hunger problems in some areas (Kang and Priyadarshan, 2007). The advances in production contributed to nutrition and better health in many areas, yet persistent poverty, especially in sub-Saharan Africa, continues to prevent food from reaching many who most need this basic resource. The inequities in distribution of food appear to be growing today, along with a skewing of the economic situation between rich and poor, North and South, all in the face of food surplus in favored areas. The current move toward biofuels, especially ethanol from maize and rapeseed in the North and from sugarcane and oil palm in the south, provides another challenge to food production and availability.

The increases in food production have been due in large part to expansion of irrigated farmlands and also to increased human productivity, based on mechanization that made the farming work load lighter. This process liberated labor to pursue other activities in the later part of the industrial revolution. Of special importance is the series of genetic advances in our major food crops that has sparked the Green Revolution, uniquely impacting rice and wheat production in Asia and Latin America and maize production in temperate regions. These advances have been coupled with large increases in chemical fertilizer application and use of chemical pesticides. The combination of these components in well-designed and efficien cropping systems has produced synergisms among the new technologies to increase food production.

Genetic improvement of crop varieties has been a continuous process since the firs people settled in permanent communities and began to extract the most desirable plants from their nearby environment (Plucknett and Smith, 1986). They saved seed of cereals and pulses and propagated cuttings from trees and vines that proved most desirable for their home diets. Often they were plants with the largest edible seeds, those that held tightly in the head or ear or pod rather than shattering and dispersing, and those with the best cooking qualities. Most of the genetic progress in improving yields of crop plants was achieved by women who found these early selections, while men were off hunting, and we have continued to fine-tun their efforts over the past centuries. Rediscovery of Mendel's principles of genetics was a key to understanding the mechanisms of crossing and development of hybrid maize. Genetic selection techniques for important self-pollinated cereals such as wheat and rice were equally successful. The contributions of the International Agricultural Research Centers and their partners in national programs were central to genetic progress in major crops during the last half of the twentieth century. Plants provide three-quarters of the calories and protein to fuel human diets, and it is valuable to explore the advances in the three major cereals, since together they contribute over half of all the energy in our food on a global basis.

Advances in Wheat

Wheat is one of our most important cereals for human consumption (see Figure 1.1, FAS, 2008), with global annual production of nearly

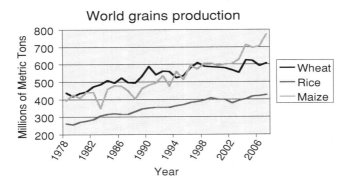

Figure 1.1. World grains production, 1978–2008 (FAS, 2008).

600 million tons (Singh and Trethowan, 2007). Half of the production is in developing countries, and much of the genetic progress can be attributed to improving adaptation to a range of water-limiting conditions through shuttle breeding carried out by the International Center for Improvement of Maize and Wheat in Mexico. Spring wheat can be grown from the equator to as far as 60° north and south latitudes and from sea level to over 3,000 m elevation. Winter wheat provides another type of adaptation to low winter temperatures and after passing through a vernalization phase can sprout early in spring and produce high yields by midsummer. Together, these wheat varieties are adapted to a wide range of ecoregions and have proved to be especially drought tolerant. Some of the highest yields have been achieved in Europe under favorable rainfall conditions, using high levels of fertilizer and growth regulators to prevent excessive vegetative growth and lodging or falling over before harvest. These systems appear to be sustainable, as long as the supply of fossil fuels, fertilizers, and chemicals needed to produce them are available. Finally, the sustainability will depend on environmental regulations and how well we are able to use these inputs without creating excessive nitrate or chemical residue loads on the environment that are detrimental to humans and other species.

Two of the irrigated areas where yields have increased in an impressive way are northwest Mexico where many of these new wheat varieties were developed, and in the Punjab of India and Pakistan where water has been available and double cropping with a summer cereal or pulse crop has been possible. Such systems are sustainable as long as soil fertility can be maintained, subject to the same limitations described for the cereal system in Europe and the availability of increasingly scarce irrigation water. In the case of the Punjab, water tables have been declining as much as 1 m/year due to intensive use of tube wells and irrigation for both winter and summer crops. At this rate, water soon becomes too costly to pump for agriculture.

Competition for water from other sectors is a critical factor. It requires about 1,000 tons of water to produce a ton of grain (Gleick, 2000). Even with the current abnormal rise in basic cereal grain prices, our economic productivity per unit of water is far below that of other industries, and it is also impossible to compete with communities for water needed for public supplies. The real advantage of agriculture in use of water is the potential to intercept rainfall over an extensive area, store this in the soil profile and use it to grow crops. Once that resource is concentrated—in a stream, reservoir, or groundwater aquifer—it is more valuable to other sectors of society. Even recreational uses and maintaining habitat for wildlife species have higher values for some human societies, at least with the current adequate levels of food production in most areas of the world. This could change as food becomes scarce and water is needed to help supply this basic human need.

Advances in Rice

Rice is another important cereal grain (see Figure 1.1, FAS, 2008), also with annual global production over 400 million tons (Virmani and Ilyas-Ahmed, 2007), and 90% grown in Asia. The crop can be grown from 35° south in Australia to 50° north in Mongolia, and from sea level to over 3,000 m elevation. Egypt and Australia have the highest levels of productivity, with average yields over 9 tons/ha. In the latter half of the past century, rice-growing area has increased almost 1.8 times, while yields per hectare have more than doubled on average, resulting in a fourfold increase in global production. Major technological advances have included breeding varieties that are insensitive to photoperiod, and thus can be planted in any month of the year where water and temperature conditions are favorable; semidwarf varieties that respond to fertilizer with more grain rather than vegetative growth; shorter maturity varieties that allow two or three crops per year if irrigation is available; and chemical nutrient and weed management to support the highly extractive practices associated with high yields and multiple crops per year.

In addition to limitations on water, one of the most bothersome issues to emerge over the last two decades with rice has been yield decline in Asia. Although a number of theories have been proposed, it now appears that nitrogen availability at the right times in the crop's life cycle is one of the principal factors (Doberman et al., 2000). Research on this critical issue continues, since rice is such an important component of the diet throughout Asia. There have been a number of concerns, including the possibility of sub-detectable effects of soil pathogens, complex questions of nutrient availability in a continuous cultivation pattern of the same crop, and other potential soil nutrient reasons for the decline. It is essential that this problem be solved for the well-being of millions of people in Asia. The potential for solving the challenge through crop rotation is an obvious route to take, yet the suitability of these lands for rice cultivation and the continuing demand for this popular food crop are overriding reasons to fin   ways to make continuous cultivation sustainable, as difficul as this may be biologically.

Advances in Maize

Maize is the most important major global cereal crop in terms of total production, nearly 700 million tons annually (see Figure 1.1, FAS, 2008). This is an important cereal in much of Africa and Southeast and East Asia, far from its origin in Guatemala and southern Mexico. As the firs cross-pollinated crop to receive major attention from plant breeders, maize has become a model for plant genetic improvement and the most important cereal grain in the United States and several other temperate countries. In addition to selection for crop yield, early efforts focused on increasing protein and oil concentrations in the grain with the hope that this would not reduce grain yields (Johnson, 2007). Development of inbred lines of maize and their heterozygous crosses became the standard for study of population genetics, testing methods, and more recently marker-assisted selection and other microbial techniques. Double-cross maize hybrids (four inbred parents) and then single-cross hybrids (two inbred parents) formed the foundation of the hybrid maize industry in the United States and a model for other countries seeking high yields and wide adaptation of new genetic combinations.

In addition to the hybrids based on inbred lines, pioneering plant breeders with maize introduced population improvement and other types of varieties that could be grown and their seed saved by farmers. This concept was built on knowledge of the pollination habits of maize; 95% of the pollen fertilizing a given ear on a plant comes from another plant—we would say the crop is 95% outcrossing. Thus, 95% of the seed that comes from a variety in the fiel is a result of crosses with some other series of plants in that fiel or nearby. By mixing a population of plants with similar characteristics, including grain color and crop maturity, one can harvest seed from the "best" plants in the fiel and assure that the large majorities are hybrids between two parents in that same field By starting with a relatively diverse and highly productive variety (population or synthetic variety), it is possible for the farmer to select an improved variety that will be even better for his or her specifi farm conditions. It is important to do this selection of plants in the fiel , choosing those individuals that stand up well, have insect and pathogen tolerance, and are well adapted to the cropping system. Those farmers who select ears, after they are in storage, often have taken the largest ones in hopes that this will increase yields, only to fin that these came from the latest maturing and often overly tall plants, both negative attributes not visible in the ears. The development of farmer varieties avoids the need to purchase hybrid seed anew each season, and the strategy has been used with success in a number of developing countries to increase maize production.

A relatively new development in crop improvement has been the introduction of transgenic hybrids of maize, rather mistakenly called GMOs, or genetically modifie organisms. This is a misnomer because all of our domesticated crops have been genetically changed since the firs farmers chose plants with larger seeds or good food quality to increase and plant near their homes. Transgenic hybrids are made from lines that have one or more genes introduced from another line or species through molecular transfer techniques. They have been used to confer resistance to specifi insect pests or specifi herbicides. This is an expensive but highly effective way to incorporate special traits into the genetic package, the seed, which can simplify management and possibly reduce input costs. One likely downside to the technology beyond its cost is the potential for developing insects or weeds with genetic resistance to a single type of control when the new hybrids are widely deployed. Farmers are currently urged to use diversity in their planting of these new hybrids and to combine them with other control strategies so that the technology will last longer.

Chemical Fertilizers

The introduction of manufactured chemical fertilizers, especially during and after the Second World War, brought more convenience to agriculture and spurred the move toward a domination mentality in farmers about how to supply needed crop nutrients in food production. In a few short decades, we abandoned many of the ecological principles, including diverse crop rotations and crop–animal integration, which had been the foundation for much of agriculture and practiced by farmers since before biblical times. In the pursuit of ever-higher yields to increase income and feed a growing human population, with both cereals and increased protein from livestock, we moved away from systems that worked with nature's cycles and resources toward systems that created large and homogeneous field and attempted to dominate the production environment.

This strategy of supplying needed nutrients to high-demand crops spurred a new industry of chemical fertilizers, especially to supply the major nutrients—nitrogen, phosphorus, and potassium—and starter fertilizers that helped the crop plants in the initial stages of establishment and growth. Fertilizer recommendations were designed to replace those nutrients extracted through the harvested crop, and higher levels just to be sure there was an adequate supply for good rainfall years and maximum crop yields. Yet, too often the early strategies did not include careful nutrient budgeting that took into account other sources of nutrients, such as those in irrigation water, those left over from the previous year, and those available from crop residues that break down and supply needed elements to the next crops. Starter fertilizers help the crop in early stages to mobilize scarce nutrients, especially phosphorus in cold climates, and plants appear to be greener and healthier. Yet, most often these early visible signs of plant vigor are not reflecte in yield differences at harvest; thus, the inputs are not economically sound. And overapplication of soluble nutrients such as nitrogen can lead to loss from the fiel through leaching down the soil profil into groundwater and surface runoff through erosion that reaches streams and lakes. This is a waste of economic resources by the farmer as well as an expense to the environment and to society that has to fin ways and resources to remediate the problem.

Good farm managers today are astute nutrient managers who take advantage of all available information from research as well as from their field and personal experiences. They carefully take into account all sources of nutrients in a complete budget—for example, nitrogen from the previous crop and from soil organic matter, that from irrigation water or anticipated from winter snows and spring rains, that from legume or grass cover crops, all before deciding how much of that essential nutrient to apply. There is technical potential to sample soils and check yields across the fiel to see where nutrients are most needed and where they are likely to give the highest crop response.

In spite of the best practices in conventional agriculture, we apply increasingly higher levels of nutrients per unit of harvested yield, and most research on chemical approaches to increase nutrient efficien y is becoming more expensive per unit of yield gain. In chemical systems, we have clearly reached the point of diminishing returns to research, at least on the major cereal crops using current research methods.

Chemical Pesticides

Introduction of synthetic chemical pesticides has been another boon to the convenience of farming. Similar to the nutrient situation, in-creased use of insecticides, herbicides, fungicides, nematicides, and other chemical controls removed the perceived need for crop rotations that had previously contributed to pest management. These products gave quick results by killing insects and weeds within hours, and were seen as a modern solution that would assure protection against pests that had plagued agriculture for centuries. In fact, some of the chemi-cal methods such as weed control through herbicides helped to reduce cultivation and soil erosion. The publicity stimulated by *Silent Spring*, a landmark book by Rachel Carson (1962), created awareness of the unintended side effects, or emergent properties, of the wide application of DDT and other chlorinated hydrocarbons and other pesticides. From the creation of the United States Environmental Protection Agency up to the present, the testing and licensing requirements for new chemicals to be used in agriculture have become much more stringent, and most current chemical pesticides are considered by many to be safe. Others maintain that we should be more cautious about the wide deployment of chemicals whose long-term effects may be hard to determine, and that the precautionary principle should guide our decisions. It is rather surprising that many biologists did not realize that wide use of any spe-cifi chemical would result in weeds, insects, and other pests that are resistant to that chemical. The current use of Roundup-Ready© corn and soybeans in a 2-year rotation, with application of the same chemical both years, will further pressure weed species to evolve with resistance. Today, there are more than 1,000 species of insects and weeds that have been shown to have resistance in the fiel to one or more chemicals (Miller, 2004).

One striking change in agriculture that is partly a result of the chemical fertilizer and pesticide revolution has been the simplificatio of farming systems to continuous cultivation of a single species or a simple 2-year rotation, for example, maize and soybeans in the United States Midwest. A corresponding change in livestock production has been the consolidation and concentration of beef, dairy, swine, and poultry into large confinemen units that are often separated in management from crop farming. Confinemen grain feeding has reduced the demand for forages; thus, there are fewer hectares of pasture and alfalfa, useful in traditional fiel rotations and in building soil fertility and controlling pests. A side effect of confinemen is the creation of a manure problem, the conversion of a high-quality production input into a waste material that needs to be disposed. A number of creative solutions are being implemented such as composting and generation of methane gas from digesters in feedlots. These provide relatively efficien ways to cycle nutrients back into the production process and solve the expensive waste disposal problem.

Postharvest Loss

The Food and Agriculture Organization (FAO) has long recognized the potential for improving food availability that can come with reductions in postharvest loss (FAO, 1981). Postharvest losses have been estimated to be about 21% of the total food in our current supply chain (Niranjan and Shilton, 1994).

Postharvest losses come from every stage after the food has been removed from the plant (ocean or animal), from harvesting, handling, storage, and transport. The reasons and amounts of losses vary greatly and depend on the crop or food and its location. The firs means to reduce losses is proper cultivation to prevent any disease or pest problems. Then it is important to plan appropriately for the harvest, planting what is needed (with some overage), harvesting in the correct conditions, and harvesting at the right time. Harvesting should also have minimal physical impacts on the product to prevent accelerated physiological deterioration. Further control measures after the fiel that reduce losses are proper storage conditions (temperature and humidity), which for some foods can mean immediate cooling.

It is assumed that reducing high postharvest losses requires technological advances. On the contrary, access to the correct means to prevent the losses is the largest hurdle for loss reduction. Limited access to the appropriate means of loss control can cause significan shortfalls. For example, postharvest losses in small-scale fisherie can be among the highest for all commodities (UN Atlas of the Oceans, http://www.oceansatlas.org/html/moreinfo.jsp, accessed June 11, 2008) often due to limited refrigeration and freezer facilities, especially onboard the fishin boats.

Organic Farming

Organic farming is often put forward as an economically viable way of sustainable food production. There has been an annual increase of over 20% in the organic food market in the United States for the past two decades (Om Organics, 2008). This is a significan achievement and one of the only major growth areas in an industry that generally considers food markets as inelastic, only growing with population and as subsequent demand increases.

Organic farmers manage nutrients without application of chemical fertilizer, using a combination of crop rotation of species with different nutrient needs, application of animal manure and/or compost, soil-building cover crops in the sequence, and calculating a careful nutrient budget to assess crop removal as well as potential for building soil fertility over time. Good managers can reduce costs of buying and applying excess nutrients, and at the same time avoid contributing to environmental problems due to nutrient loss from farm fields

Organic farms are known for use of manure and compost and frequent integration of crops and animals on the same farms to make nutrient cycling more efficien (USDA, 2007). Although there may be excessive cultivation and resulting soil erosion on some organic farms, this approach to agriculture does reduce chemical load in the environment and engenders efficien use of on-farm, renewable production resources.

Another impact of technology has been an increasing industrialization of organic agriculture. The attractiveness of the organic segment of the food industry is increasingly recognized by those in the global agriculture and food sectors, and today over half of all organic food is marketed through large corporate supermarkets rather than small, locally owned, specialized organic food stores. There is concern among many of the founders of the organic food movement that much of the social intent of the original concept is lost, and they are searching for a more restrictive type of certificatio that would emphasize more than production methods: fair treatment of farm labor, emphasis on local foods, and distribution of benefit to a wider group of citizens.

In contrast to the potentials reported by most authors, there is still debate about potential global production from organic agriculture (Sustainable Food News, 2007). The FAO stated that organic agriculture should be used and promoted for its wholesome and nutritious value as well as the growing income it is providing for developed and developing countries; however, with current yields and land use, it may not be able to feed the 6 billion people on the planet today and the potentially 9 billion in 2050 without judicious use of chemical fertilizers (Sustainable Food News, 2007). We challenge that conclusion, based on personal observations, as the best organic farmers in the Midwest consistently produce crop yields as high if not higher than county averages.

Legislation and Supports

Technologies in agriculture, especially in the United States and the European Union, have been highly successful in raising productivity and increasing production far beyond internal needs and local markets. This has led to major exports of cereals from these regions and growth and later consolidation of grain marketing into a few major corporations.

Although often held up as a model for the success of free market economies and touted as an industry that has benefite farmers in the North as well as food-deficien countries in the South, in fact, the agricultural export industry has brought focused benefit to the larger farm operations and to those supplying industrial agriculture. This includes corporations supplying inputs, commodity traders and exporters. While benefitin these larger operations, the agricultural export industry has often suppressed successful production and skewed markets in many developing countries. The North American Free Trade Agreement is the latest example of this activity of the free market agricultural economy. Far from creating a level-playing fiel , the legislation and supports in the two major food export countries in North American have favored large-scale operations and funded exports in a number of ways, while small farmers in Mexico have been forced out of farming.

One of the most prominent liberal economists in agriculture, and advisor to presidents from both major parties, is Willard Cochrane from Minnesota. According to Cochrane, long-term attempts to stabilize production and farm incomes through farm programs in the United States have surely done some good by providing subsidies for maintaining prices during hard times and promoting export of food grains (Cochrane, 2003). But they have failed to provide long-term stability, and the result over the course of more than 70 years has been a continuing consolidation of ownership, exit of farmers and farm families, and decline of rural communities. Cochrane further concludes that large regional cooperative projects such as the North American Free Trade Agreement have contributed to greater, rather than lesser, inequities in incomes and success, and are especially destructive for small and family farms in the United States as well as in the other two partner countries.

Another respected agricultural economist in the land grant system, John Ikerd from University of Missouri, maintains that the path toward long-term security in the United States food system is through sustainable agriculture (Ikerd, 2006). In order to solve the negative and unexpected environmental consequences of the current industrial model of agriculture, it is essential to reduce the contamination of waterways and aquifers from pesticide residues and chemical fertilizer nutrients. These residues come from the more than 1.1 billion pounds of pesticide applied annually (Kiely et al., 2004) and over 12 million tons of nitrogen applied annually in the United States (ERS, 2007). The residues in the environment from these chemicals are indicators of the decline in ecological sustainability of present production systems. As Ikerd asserts, recognition of the environmental impacts of conventional agriculture has led to greater scrutiny of the economic and social sustainability of these same systems.

Consolidation has partly been a result of farm support programs (Ikerd, 2008), since payments have been coupled to production, allowing the larger operators to acquire more capital, which is then put into land purchase. With small profi margins on conventional commodity crops, the common wisdom in the western Corn Belt is that a family must farm at least 1,000 acres to earn enough net income to support an adequate lifestyle. As with any conventional wisdom, this represents an average farm size—some farm families add value to products on farm and do well on smaller farms, for example, those that are certifie organic and have premiums for their products. Farmers who use imagination to diversify their crops and animal enterprises reduce costs by using primarily internal resources from the farm for maintaining soil nutrition, manage pests through rotation and diversity, and direct market their products often claim that a much smaller farm is adequate.

There is a growing interest in the United States and a strong initiative in the European Union in recognizing the importance of multifunctional rural landscapes, especially as they provide a range of ecosystem services to the larger society. These have been summarized in an the book *Nature's Services: Societal Dependence on Natural Ecosystems* by Daily (1996). One of the services that is already recognized economically and traded on the futures market is carbon credits. There are also federal programs (United States) and regional programs (European Union) that reward conservation-related practices with annual subsidies. One of the clearest mandates on agricultural research in recent years in the United States that supports this new direction was the comprehensive report of the National Research Council on priorities for research in the future. The prestigious panel that assembled this report clearly identifie the importance of a multifunctional agriculture and rural community development as the foundation of viability for the rural sector in the United States (National Research Council, 2003), and asserted that agricultural landscapes are important for much more than food production. In so doing, they acknowledge the importance of families, communities, and ecosystem services. Yet we recognize that it will be years before a research establishment, as large and complex as that in the United States, one driven primarily by agricultural production and supported strongly by input and grain companies, will make any major change in direction. It is apparent that we do need a broader approach to marketing in a complex and greener future that includes more environmental concern by consumers and investors.

Consequences of Current Approaches and Paths Forward

Although the high-technology approach to agriculture has resulted in rapid increases in productivity and production, and one consequence is an increased availability of food to many, there have been some unexpected and negative consequences (Horrigan et al., 2002). The focus on crop yields and maximizing profit as a single strategy has led to economic, environmental, and social problems (Shrestha and Clements, 2003).

Keoleian and Heller (2003) summarized these consequences. They found that rapid conversion of prime farmland to urban development led to less stable and less arable land being used for agriculture resulting in increased erosion, and irrigation leading to depletion of topsoil exceeding regeneration and rate of groundwater withdrawal exceeding recharge in major agricultural regions. Losses to pests are increasing, despite use of chemicals. With a 10-fold increase in insecticide use from 1945 to 1989, there was a concurrent increase in losses from insects from 7 to 13% (Pimentel et al., 1991). More recent estimates suggest that there is a loss of 37% of all crops, globally, due to pests (insects, pathogens, and weeds), in spite of massive applications of pesticides (Pimentel and Pimentel, 2008). There is a reduction in genetic diversity, since today only 10–20 crops provide 80–90% of the world's calories (Brown, 1981). Such lack of biodiversity makes the supply more susceptible to pests and disease, leading to declining economic conditions for farmers, especially as production moves from smaller farms to larger, industrial farms. This is because the price of crops is low, yet investment is high. This means that increasingly only the larger operations are able to make both ends meet.

Thus, we have a complex global food situation, where large amounts of food are lost to pests, costs of inputs are increasing rapidly, there is competition between production for food and fuel, and great inequities exist between the North and South. While many in the world are undernourished, a number of countries in the North are seeing overconsumption and obesity as growing health challenges. These themes are expanded, as both social and economic impacts of the current food system are discussed in Chapter 6.

There is a heavy reliance on fossil fuels resulting in an imbalance in energy input and output that further burdens the system and environment. On a global scale, agriculture significant y contributes to greenhouse gas (GHG) emissions, producing between 17 and 32% of all global human-induced GHG emissions (Bellarby et al., 2008). This includes land and livestock direct emissions, fossil fuel use in farm operations, production of agrochemicals, and energy costs of the overall food system. The largest contributors are conversion of land to agriculture and direct emissions from land and animals. For example, excess use of fertilizers releases nitrous oxide, a GHG, and fertilizer release of nitrous oxide is the single largest contributor to GHG in agriculture (Bellarby et al., 2008). This does not even include GHG contribution for the production of fertilizers. Livestock is the largest user of land, using one-third of the world's arable land due to the shift away from grazing to the growth of livestock feed crops (Compassion in World Farming, 2008). This has caused major deforestation and other native vegetation destruction. Such conversion is a net release of carbon, given that native vegetation, such as trees, stores more carbon than crops like soy (Bellarby et al., 2008).

Bellarby et al. (2008) suggest that the significan environmental impacts of agricultural production can be mitigated with a number of approaches. These focus on shifting farming practices to alternatives that provide carbon sequestration rather than emission such as improved cropland management (such as avoiding bare fallow/using cover crops and appropriate fertilizer use), grazing-land management, and restoration of organic soils as carbon sinks. Further, since meat production is inefficien in its delivery of energy to the human food chain and at the same time is a significan source of GHG emissions, a reduction of meat production and consumption could provide major improvements. It has been suggested that a reduction in the size of the livestock industry is the simplest, quickest and probably the only effective method of cutting GHGs from animal production to the extent that is necessary to limit the future increase in global warming (Compassion in World Farming, 2008).

The challenges in agriculture and the global food system are many and complex. It is certain that neither single nor simple solutions can be found to resolve all the problems outlined above, and any changes in strategy must take into account the overriding need to provide adequate food for a growing world population. The last several decades have demonstrated that impressive increases in food production can be achieved, while we have also learned that there are often unanticipated consequences, or emergent properties, of any widespread application of new technologies. What we need is balance, careful study of available alternatives, and assessment of the multiple economic, environmental, and social impacts of proposed changes. Most important of all, we need to accept that there are limits to growth, and an essential focus for future technologies in agriculture must be to improve the qualitative dimensions of farming systems and rural life, and not just increased yields and total production.

References

Bellarby, J., B. Foereid, A. Hastings, and P. Smith. 2008. *Cool Farming: Climate Impacts of Agriculture and Mitigating Potential.* Amsterdam, the Netherlands: Greenpeace International.

Brown, L.R. 2008. *Plan B 3.0: Mobilizing to Save Civilization.* New York: W.W. Norton & Company.

Brown, N.J. 1981. "Biological diversity: The global challenge." In: *U.S. Strategy Conference on Biological Diversity.* Washington, DC: U.S. Department of State.

Carson, R. 1962. *Silent Spring.* New York: Houghton Mifflin

Cochrane, W.W. 2003. *The Curse of Agricultural Abundance: A Sustainable Solution.* Lincoln, NE: University of Nebraska Press.

Compassion in World Farming. 2008. *Global Warning: Climate Change and Farm Animal Welfare.* U.K.: Compassion in World Farming. Available from http://www.ciwf.org/publications/reports/global-warning-summary.pdf. Accessed June 3, 2008.

Daily, G.R. 1996. *Nature's Services: Societal Dependence on Natural Ecosystems.* Washington, DC: Island Press.

Doberman, A., D. Dawe, R.P. Roetter, and K.G. Cassman. 2000. Reversal of rice yield decline in a long-term continuous cropping experiment. *Agron. J.* 92:633–643.

ERS. 2007. *US Fertilizer Use and Price.* U.S. Department of Agriculture, Economic Research Service. Available from http://www.ers.usda.gov/Data/FertilizerUse/. Accessed October 2007.

FAO. 1981. *Food Loss Prevention in Perishable Crops.* Available from http://www.fao.org/docrep/S8620E/S8620E04.htm. Accessed September 16, 2008

FAO. 2007. *The State of World Fisheries and Aquaculture (SOFIA) 2006.* Rome: Food and Agriculture Organization, Fisheries Department.

FAS. 2008. *Grain: World Markets and Trade.* U.S. Department of Agriculture, Foreign Agricultural Service, FAS Circular Series FG0308. Available from http://www.fas.usda.gov/grain/circular/2008/03-08/graintoc.asp. Accessed April 2008.

Francis, C.A., R.P. Poincelot, and G. Bird, eds. 2006. *Developing and Extending Sustainable Agriculture: A New Social Contract.* Binghampton, NY: Haworth Publishers.

Gleick, P.H. 2000. *The World's Water 2000–2001: The Biennial Report on Freshwater Resources.* Washington, DC.: Island Press.

Horrigan, L., R.S. Lawrence, and P. Walker. 2002. How sustainable agriculture can address the environmental and human health harms of industrial agriculture. *Environ. Health Perspect.* 110(5):445–456.

Ikerd, J.E. 2006. *Sustainable Capitalism: A Matter of Common Sense.* Bloomfiel , CT: Kumarian Press.

Ikerd, J.E. 2008. *Crisis and Opportunity: Sustainability in American Agriculture.* Lincoln, NE: University of Nebraska Press.

Johnson, G.R. 2007. Corn breeding in the twenty-firs century. In: *Breeding Major Food Staples*, eds M.J. Kang and P.M. Priyadarshan, pp. 227–244. Ames, IA: Blackwell Publishing.

Kang, M.S., and P.M. Priyadarshan, eds. 2007. *Breeding Major Food Staples*. Ames, IA: Blackwell Publishing.

Keoleian, G.A., and M.C. Heller. 2003. "Life-cycle based sustainability indicators for assessment of the U.S. food system: Life cycle assessment in the agri-food sector." In: *Proceedings from the 4th International Conference, October 6–8, 2003*, Bygholm, Denmark.

Kiely, T., D. Donaldson, and A. Grube. 2004. *Pesticides Industry Sales and Usage 2000 and 2001 Market Estimates*. U.S. Environmental Protection Agency. Available from http://www.epa.gov/oppbead1/pestsales/. Accessed October 2007.

Miller, G.T. 2004. *Sustaining the Earth*, 6th edn. Pacifi Grove, CA: Thompson Learning, Inc.

National Research Council. 2003. *Frontiers in Agricultural Research: Food, Health, Environment, and Communities*. Washington, DC: National Academies Press.

Niranjan, K., and N.C. Shilton. 1994. "Food processing wastes—their characteristics and an assessment of processing options." In: *Environmentally Responsible Food Processing*, ed. E.L. Gaden. New York, NY: American Institute of Chemical Engineers.

Om Organics. 2008. *History of Organics*. Available from http://www.omorganics.org/page.php?pageid=82. Accessed January 24, 2008.

Pimental, D., and M.H. Pimentel, eds. 2008. *Food, Energy, and Society*, 3rd edn. Boca Raton, FL: CRC Press, Taylor and Francis Group.

Pimentel, D.L., L. McLaughlin, A. Zepp, B. Lakitan, T. Kraus, P. Kleinman, F. Vancini, W.J. Roach, E. Graap, W.S. Keeton, and G. Selig. 1991. "Environmental and ecological impacts of reducing U.S. agricultural pesticide use." In: *Handbook on Pest Mnagement in Agriculture*, ed. D. Pimental, pp. 679–718. Boca Raton, FL: CRC Press.

Plucknett, D.L., and N.J.H. Smith 1986. "Historical perspectives on multiple cropping." In: *Multiple Cropping Systems*, ed. C.A. Francis, pp. 20–39. New York: Macmillan Publishing Co.

Shrestha, A., and D.R. Clements. 2003. "Emerging trends in cropping systems research." In: *Cropping Systems: Trends and Advances*, ed. A. Shrestha, pp. 1–14. Binghampton, NY: Haworth Press.

Singh, R.P., and R. Trethowan. 2007. "Breeding spring wheat for irrigated and rainfed production systems of the developing world." In: *Breeding Major Food Staples*, eds M.J. Kang and P.M. Priyadarshan, pp. 109–140. Ames, IA: Blackwell Publishing.

Sustainable Food News. 2007. *Organic Agriculture Cannot Feed the World: FAO Says Chemical Fertilizers Needed to Ensure Food Security*. Available from http://www.sustainablefoodnews.com/story.php?news_id=3056. Accessed December 10, 2007.

Tilman, D., K.G. Cassman, P.A. Maatson, R. Rosamond-Taylor, and S. Polasky. 2002. Agricultural sustainability and intensive production practices. *Nature* 418:671–677.

USDA. 2007. *Organic Production.* U.S. Department of Agriculture, Economic Research Service. Available from http://www.ers.usda.gov/Data/Organic/. Accessed October 2007.

Virmani, S.S., and M. Ilyas-Ahmed. 2007. "Rice breeding for sustainable production." In: *Breeding Major Food Staples*, eds M.J. Kang and P.M. Priyadarshan, pp. 141–192. Ames, IA: Blackwell Publishing.

Chapter 2

Food Processing and Food Waste

Tran Thi My Dieu

Introduction

The food processing industrial sector is a large and rapidly growing industry and plays an important role in economic development across the world. There are several reasons attributed to the important role of food processing. First, it processes agricultural raw materials (such as grain, maize, cassava, sugarcane, coffee beans, fruits, and vegetables) to provide new products for human life and simultaneously helps to raise the income of farmers, who used to be poorer than nonfarmers. Secondly, food processing factories create jobs (and in some cases more than other industrial sectors as it is relatively labor-intensive). Finally, the food processing industrial sector adds value to agricultural products before exportation, and thus it contributes to economic development (Dieu, 2003).

The main products of food processing consist of cereals, fruits and vegetable, roots and tubers, sugar crops, and milk. Values of agricultural gross domestic product (GDP) compared to per capita GDP of different countries are presented in Table 2.1, which prove the role of the agricultural and related processing industry in economic development of different developed, developing, and less developing countries in the world.

Industrial production activities, however, have impacts on the natural environment through the entire cycle of raw materials exploration and extraction, transformation into products, and use and disposal of products by the fina consumers. When considering the scope of food processing, the main (direct) impacts are from waste generation, water

Table 2.1.   General data of population and agricultural GDP of different countries in the world[a]

Items	Australia	Bangladesh	Cambodia	China	India	Indonesia
Population[b]	19,913	1,984	14,482	1,320,892	1,081,229	222,611
Per capita GDP[c]	22,303	47	309	1,441	538	886
Agricultural GDP/average population[c]	20,826	18	148	241	201	325

Items	Japan	Korea Republic	Malaysia	Thailand	United States	Vietnam
Population[b]	127,800	47,951	24,876	63,465	297,043	82,481
Per capita GDP[c]	39,184	12,793	4,277	2,359	36,352	499
Agricultural GDP/average population[c]	16,714	6,973	2,359	413	27,651	159

[a] Data of 2004.
[b] 1,000.
[c] US dollar.
Source: FAO Statistical Yearbook, 2005–2006.

use, and energy use. The discussion below will include water use along with waste as they are often closely related.

Waste

Food waste is significan through the supply chain. It is estimated that the losses primarily come from the farm due to spoilage (about 21% of supply) and processing (about 7% of supply) (Niranjan and Shilton, 1994). The main types of processing waste are solid, water, and emissions (dust, volatile organic compounds, and odor) (Niranjan and Shilton, 1994).

Most industrial activities necessarily generate wastes and/or by-products. For instance, typical cheese production separates the milk so that part of the milk is not used and is a *by-product* (Erkman, 1997). Neither enterprises nor consumers want to store wastes in their own yards. Therefore, they have to fin places and methods to get rid of it, and frequently the natural environment is serving as the recipient of all kinds of wastes, including industrial wastes with high toxicity and loading of contaminants. This explains why industrial wastes (including wastewater, solid wastes, and air pollution) are one of the major causes of severe environmental deterioration in the world. Besides the most notable polluters in the heavy industrial sector (such as power plants, iron and steel mills, pulp and paper factories, cement plants, chemical plants, and fertilizer companies), food processing companies are considered to be one of the major polluters in the light industrial sector, involving companies in the food processing, textile and dyeing, electroplating, and leather tanning subsectors (Fryer, 1995; Frijns et al., 2000). Within food processing, the largest producers of waste come from milk, cocoa/chocolate/sugar confections, brewing/distillation, and meat processing (Niranjan and Shilton, 1994).

It is known that a lot of waste generated from industrial activities may still contain valuable components, but these are wasted due to a variety of reasons:

– Enterprises do not know how to recover these valuable components.
– The economic benefi of reuse and recovery is low.
– Due to the absence of waste exchange or a waste exchange information center, nobody knows who might be interested in reusing residue components.
– There is an absence of legislation and incentives to encourage reuse and recycling activities and resource conservation.
– Low costs and fine for waste discharge/disposal make enterprises uninterested in the amount of wastes generated.

Waste Generation

Waste generation from food processing includes wastewater, solid waste, and polluted air. Wastewater contains biological materials and dissolved solids (organic and inorganic); solid waste includes food and packaging; and emissions include dust, volatile organic compounds, and odor (Nimi and Gimenez-Mitsotakis, 1994). As mentioned previously, the main food processing waste comes from milk, cocoa/chocolate/sugar confections, brewing/distillation, and meat processing (Niranjan and Shilton, 1994).

For example, it was estimated that in 1994 15 million metric tons of whey was produced as a by-product of cheese production (Clanton et al., 1994). Half of that gets used for human and animal consumption;

the rest is waste and must be treated before releasing into the environment (usually waterways).

Refine y sugar production usually generates molasses, activated carbon, filte aid powder, and resin. Tapioca starch manufacturing industries release cassava hard roots and wood shells, fibrou residues, and pulp, both in solid and liquid wastes (Dieu, 2003).

Breweries can achieve an effluen discharge of approximately 3–5 m^3/ ton of sold beer (exclusive of cooling waters), while solid wastes for disposal include grit, weed, seed, and grain of less than 2.2 mm in diameter, removed when grain is cleaned, spent grain and yeast, spent hops, broken bottles, or bottles that cannot be recycled to the process (World Bank, 1998).

Waste stream of meat products manufacturing industries can contain blood, meat and fatty tissue, meat extracts, paunch contents, bedding, manure, hair, dirt, contaminated cooling water losses from rendering, curing, and pickling solutions, preservatives, caustic, or alkaline detergents (Carawan and Pilkington, 1986).

Wastewater

Wastewater generation and characteristics from different food production processes can be referred from a study from the World Health Organization (WHO) (1993), as summarized in Table 2.2. Values presented in Table 2.2 show that among food production processes, yeast manufacturing (150 m^3 wastewater/ton of product), breaded and frozen shrimp processing (116 and 115 m^3 wastewater/ton of product), sugarcane distilleries (113 m^3 wastewater/ton anhydrous alcohol), caulifl wer processing (89.4 m^3 wastewater/ton raw material), and grapefruit canning (72.1 m^3 wastewater/ton raw material) generate high amounts of wastewater.

Water is a raw material in the meatpacking industry and originates from animal holding pens, slaughtering, cutting, meat processing, secondary manufacturing (by-product operations), and cleanup operations (Carawan and Pilkington, 1986). As presented in Table 2.2, a simple slaughterhouse, which conducts very few of the by-product operations, generates less wastewater (about 5.3 m^3/ton LWK) than a complex slaughterhouse (about 7.4 m^3/ton LWK). Pollutants of holding pens are derived from manure and urine, feed, dirt borne by livestock, and

Table 2.2. Wastewater generation from food processing industry

Process	Unit (U)	Volume of WW (m³/U)	BOD₅ (kg/U)	TSS (kg/U)	Total N (kg/U)	Total P (kg/U)	Other pollutants Name	kg/U
Slaughtering, preparing and preserving meat								
Simple slaughterhouses								
With blood recovery	ton LWK	5.3	6.0	5.60	0.7	0.05	Oil	2.1
Without blood recovery	ton LWK	5.3	10.0	8.00	0.7	0.05	Oil	4.0
Complex slaughterhouses	ton LWK	7.4	10.9	9.60	0.84	0.33	Oil	5.9
Packing houses								
Low processing	ton LWK	7.8	8.1	5.90	0.53	0.13	Oil	3.0
High processing	ton LWK	12.5	16.1	10.5	1.30	0.40	Oil	9.0
Rendering plants	ton LWK	3.3	2.15	1.13	0.48	0.04	Oil	0.72
Poultry processing								
With blood recovery	1,000 birds	37.5	11.9	12.7			Oil	5.6
Without blood recovery		37.5	17.0	12.7			Oil	5.6
Manufacturing of dairy products								
Receiving station								
Cans	ton of product	0.68	0.46	0.03	0.49	0.11		
Bulk	ton of product	0.08	0.17	0.03	0.06	0.013		
Fluid products	ton of product	3.1	3.21	1.50	0.31	0.68		
Cultured products	ton of product	3.9	3.21	1.50	0.31	0.68		
Butter	ton of product	2.6	1.1	0.40	1.95	0.42		
Cottage cheese with whey recovery	ton of product	7.7	21.7	3.40				
Natural cheese with whey recovery	ton of product	2.3	2.2	0.20	1.56	0.34		
Ice cream	ton of product	3.0	10.9	1.50				

(*Continued*)

Table 2.2. (*Continued*)

Process	Unit (U)	Volume of WW (m³/U)	BOD$_5$ (kg/U)	TSS (kg/U)	Total N (kg/U)	Total P (kg/U)	Other pollutants Name	kg/U
Condensed milk	ton of product	2.0	6.7	0.83	0.39	0.08		
Powder production								
Spray drying	ton of product	0.7	22.4		1.30	0.28		
Roller drying	ton of product	0.8	26.8		1.56	0.34		
Canning and preserving of fruits and vegetables								
Fruits processing								
Apricots	ton raw material	29.1	15.4	4.25				
Apple								
All products	ton raw material	3.7	5.0	0.5				
All products except juice	ton raw material	5.4	6.4	0.8				
Juice	ton raw material	2.9	2.0	0.3				
Can berries	ton raw material	5.8	2.8	0.6				
Citrus	ton raw material	10.1	3.2	1.3				
Sweet cherries	ton raw material	7.8	9.6	0.6				
Sour cherries	ton raw material	12.0	17.2	1.0				
Brine cherries	ton raw material	19.9	21.7	1.4				
Cranberries	ton raw material	12.3	10.0	1.4				
Dried fruit	ton raw material	13.3	12.4	1.9				
Grapefruit								
Canning	ton raw material	72.1	10.7	1.2				
Pressing	ton raw material	1.6	1.9	0.4				

Olives	ton raw material	38.1	43.7	7.5
Peaches				
Canned	ton raw material	13.0	14.0	2.3
Frozen	ton raw material	5.4	11.7	1.8
Pears	ton raw material	11.8	21.2	3.2
Pickles				
Fresh packed	ton raw material	8.5	9.5	1.9
Process packed	ton raw material	9.6	18.3	3.3
Salting stations	ton raw material	1.1	8.0	0.4
Pineapples	ton raw material	13.0	10.3	2.7
Plums	ton raw material	5.0	4.1	0.3
Raisins	ton raw material	2.8	6.0	1.6
Strawberries	ton raw material	13.1	5.3	1.4
Tomatoes				
Peeled	ton raw material	8.9	4.1	6.1
Products	ton raw material	4.7	1.3	2.7
Vegetable processing				
Asparagus	ton raw material	68.8	2.1	3.4
Beets	ton raw material	5.0	19.7	3.9
Broccoli	ton raw material	45.6	9.8	5.6
Brussels sprouts	ton raw material	36.3	3.4	10.8
Carrots	ton raw material	12.1	19.5	12.0
Caulifl wer	ton raw material	89.4	5.2	2.7
Corn				
Canned	ton raw material	4.5	14.4	6.7
Frozen	ton raw material	13.3	20.2	5.6

(Continued)

Table 2.2. (Continued)

Process	Unit (U)	Volume of WW (m³/U)	BOD₅ (kg/U)	TSS (kg/U)	Total N (kg/U)	Total P (kg/U)	Other pollutants Name	kg/U
Dehydrated								
Onion and garlic	ton raw material	19.9	6.5	5.9				
Vegetables	ton raw material	22.1	7.9	5.6				
Dry beans	ton raw material	18.0	15.3	4.4				
Lima beans	ton raw material	27.1	13.9	10.3				
Mushrooms	ton raw material	22.4	8.7	4.8				
Onions canned	ton raw material	23.0	22.6	9.3				
Peas	ton raw material							
Canned	ton raw material	19.7	22.1	5.4				
Frozen	ton raw material	14.5	18.3	4.9				
Pimentos	ton raw material	28.8	27.2	2.9				
Potatoes								
All products	ton raw material	10.3	18.1	15.9				
Frozen products	ton raw material	11.3	22.9	19.4				
Dehydrated products	ton raw material	8.8	11.0	8.6				
Sauerkraut								
Canning	ton raw material	3.5	3.5	0.6				
Cutting	ton raw material	0.43	1.2	0.2				
Snap beans								
Canned	ton raw material	15.4	3.1	2.0				
Frozen	ton raw material	19.9	6.0	3.0				

	Unit					
Spinach						
Canned	ton raw material	37.6	8.2	6.5		
Frozen	ton raw material	29.2	4.8	2.0		
Squash	ton raw material	5.6	16.8	2.3		
Sweet potatoes	ton raw material	4.1	30.1	11.5		
Canning, preserving, and processing of fish, crustacean, and similar foods						
Fish and seafood processing industry						
Catfis processing	ton of product	24	7.3	9.4	0.65	Oil 4.7
Blue crab						
Convent processing	ton of product	1.2	5.2	0.74	1.00	Oil 0.25
Mechanized processing	ton of product	38	22.5	12	3.7	Oil 5.6
Shrimp						
Breaded	ton of product	116	84	93	5.9	Oil 20
Canned	ton of product	52	82	43	9.5	Oil 31
Frozen	ton of product	115	120	220	10	Oil 29
Tuna	ton of product	25	13.4	10.4	2.1	Oil 7.4
Clam						
Hand shucked	ton of product	4.6	5.1	10.2		Oil 0.14
Mechanized	ton of product	19.5	18.7	6.3		Oil 0.46
Fish meal						
With soluble	ton of product	35	3	0.9		Oil 0.56
Without soluble	ton of product	1.9	62.2	34.8		Oil 22.8
Salmon						
Mechanical butchering	ton of product	18.5	50.8	20.3		Oil 6.5
Hand butchering	ton of product	4.0	2.1	1.2		Oil 1.5
Sardine	ton of product	8.7	9.2	5.4		Oil 1.7
Herring fille	ton of product	7.0	32.2	20.9		Oil 6.5
Oyster/steam cleaned	ton of product	98	61.2	155		Oil 1.5

(Continued)

Table 2.2. (Continued)

Process	Unit (U)	Volume of WW (m³/U)	BOD₅ (kg/U)	TSS (kg/U)	Total N (kg/U)	Total P (kg/U)	Other pollutants Name	kg/U
Manufacturing of vegetable animal oils and fats								
Olive oil expression								
Pressing	ton of product	5	210	325				
	ton of olives	1	42	65				
Centrifuging	ton of product	7	95	455				
	ton of olives	1.4	19	91				
Edible oil refinin								
Edible fats and oils								
General	ton of product	6.8	24.9	24.6			Oil	28.1
Corn oil	ton of product	1.85	0.3	0.35				
Olive oil	ton of product	57.5	12.9	16.4			Oil	6.5
Grain mill products								
Corn								
Wet milling	ton of product	22.4	7.3	5.2				
Dry milling	ton of product	0.7	1.1	1.6				
Wheat								
Normal milling	ton of product	No effluen						
Bulgur milling	ton of product	0.29	0.11	0.1				
Rice								
Normal milling	ton of product	No efflúen						
Parboiled milling	ton of product	1.5	1.8	0.07				
Manufacturing of bakery products								
Bread	ton of product	0.11		0.004				
Rusk	ton of product	0.11		0.004				

32

	Unit							
Dry pastry	ton of product	0.7		0.005				
Wet pastry	ton of product	9		0.05				
Sugar factories and refineries								
Beet sugar production	ton of product	23	20	75				
Cane sugar production	ton of product	3–48	2.9	6.3				
Manufacture of food products not elsewhere classified								
Specialty food industry								
Prepared dinners	ton of product	12	17	14	0.44	0.19	Oil	15
Frozen bakery product	ton of product	11	23	14	0.3	0.08	Oil	11
Dressing, sauces, and spreads	ton of product	2.8	7.5	3.5	0.04	0.03	Oil	5.7
Meat specialties	ton of product	10	10	6.1	0.57	0.1	Oil	4
Canned soups and baby foods	ton of product	22	12	7.6	0.47	0.18	Oil	2.4
Tomato–cheese–starch combinations	ton of product	2.9	7.2	6.0	0.23	0.28	Oil	4.7
Sauced vegetables	ton of product	8.5	25	21	1.1	0.33		
Sweet syrups, jams, and jellies	ton of product	2.4	5.1	1	0.04	0.05	Oil	0.6
Chinese and Mexican foods	ton of product	14	6.9	2.8	0.28	0.14	Oil	3
Breaded frozen products	ton of product	48	26	26	2.6	0.35		
Egg breaking								
US facilities	ton of product	10.3	33					
Dutch facilities	ton of eggs	7.9	12.4					
	Dozen eggs	0.0039	0.006					
Wheat, starch gluten	ton of product	9.9	94	81	3.7	1		
Starch and glucose	ton of product	33	13.4	9.7				
Yeast manufacturing	ton of product	150	1,125	18.7	127		SO_4	337

(Continued)

Table 2.2. (*Continued*)

Process	Unit (U)	Volume of WW (m³/U)	BOD₅ (kg/U)	TSS (kg/U)	Total N (kg/U)	Total P (kg/U)	Other pollutants Name	kg/U
Beverage industries								
Distillery, rectifying, and blending spirits								
Grain distilleries	ton anhydrous alcohol	63	216	257				
Molasses distilleries		63	220	300				
Sugarcane distilleries		113	426					
Wine distilleries		36	210	75				
Wine production	ton of grapes	2	1.6	0.3				
Malt liquors and malt								
Beer manufacturing	ton of barley	7.3	5	0.85				
Malting and brewing								
New large plant	m³ of beer	5.4	10.5	3.9				
Old large plant	m³ of beer	11	18.8	7.3				
Soft drinks								
Major plant with syrup preparation	m³ of product	12.8	3.1	4.3				
Franchise plant without syrup preparation								
Bottled	m³ of product	4.3	2.1	0.7				
Canned	m³ of product	2.0	0.8	0.3				

LWK is an abbreviation for the live weight of the animals killed. The average weight of cattle is 430 kg, of calf 97 kg, of hog 120 kg, and of lamb 52 kg. The edible meat is about 60% of LWK. If a waste load factor does not appear in the appropriate place, this often (but not always) means that its value is either small or zero.
Source: WHO (1993).

sanitizers and cleaning agents that may be used in pen washdown (Hrudey, 1984). The major pollutant from the slaughtering operations is blood (Higgins, 1995). Meat processing covers several operations such as curing, pickling, smoking, cooking, and canning. These generate raw waste load from blood, tissue, and fats, which reach the sewer during cleanup (Hrudey, 1984).

Dairy food plant wastes are generally dilutions of milk or milk products, together with detergents, sanitizers, lubricants, chemicals from boiler and water treatment, washings from tank trucks, and domestic wastes. The processes and other sources of wastes that have a significant effect on wastewater from all dairy plant operations include the following (Marshall and Harper, 1984):

- Rinsing and washing of bulk tanks or cans in receiving operations
- Rinsing of residual product remaining in or on the surfaces of all pipelines, pumps, tanks, vats, processing equipment, fillin machines, etc.
- Washing of all processing equipment
- Water–milk solids mixture discharged to the drain during start-up, product changeover and shutdown of pasteurizers, heat exchangers, separators, clarifiers and evaporators
- Carryover of droplets of milk into the tail-waters of vacuum reactors, pasteurizers, and evaporators
- Sludge discharged from clarifier
- Fines from cheese and casein operations
- Spills and leaks due to improper equipment operation and maintenance, overfl ws, freezing-on, and incorrect handling
- Waste of unwanted by-products or spoiled materials
- Loss in packaging operations through equipment breakdowns and broken packages
- Product returns
- Lubricants from equipment, casers, stackers, and conveyors
- Washings from the outsides of tank trucks including dirt, stones, and farm debris
- Dust from coal and wood fuel and spills of fuel oil
- Powder deposited from discharges from driers
- Ash from boilers
- Water and boilers treatment chemicals
- Chemicals from the regeneration of ion exchanger resins.

Major wastewater loads from fruit and vegetable processing industries are usually generated by washing, fluming trimming, peeling, balancing, can washing, cooling, and plant cleanup. Wastewater fl ws from fruit and vegetable plants are often intermittent and vary greatly, depending on in-plant operations (Harrison et al., 1984).

Vegetable oil processing wastewater generated during oil washing and neutralization may have a high content of organic material, suspended solids, organic nitrogen, oil and fat, and may contain pesticide residues from the treatment of the raw materials (International Finance Corporation, IFC, 2007a). For production of oil based in palm oil fruit, wastewater volumes can often be limited to 3–5 m^3/ton of feedstock (European Commission, 2005).

High consumption of good quality water is characteristic of beer brewing. The pollutant load of brewery effluent is primarily composed of organic material from process activities (IFC, 2007b).

Fish processing requires large amounts of water, primary for washing and cleaning purposes, but also as media for storage and refrigeration of fis products before and during processing. Detergents including acid, alkaline, and neutral detergents and disinfectants including chlorine compounds, hydrogen peroxide, and formaldehyde (IFC, 2007c) may also be present in the wastewater after cleaning.

Food waste's environmental impact is primarily from polluted water with high biological oxygen demand (BOD), which results in growth of undesirable microorganisms (Niranjan and Shilton, 1994). The growth of the tapioca processing industry has resulted in heavy water pollution as it generates large amounts of wastewater with very high organic concentrations. Studies of Mai et al. (2001) and Oanh et al. (2001) on large-scale tapioca processing companies in Vietnam give similar characteristics of tapioca wastewater with total chemical oxygen demand (COD) in the range of 7,000–41,406 mg/L, 5-day biological oxygen demand (BOD_5) of 6,200–23,077 mg/L, and CN^- of 19–28 mg/L. Pollution of wells, springs and rivers especially in the provinces of Binh Phuoc, Ha Noi, and Dong Nai, Vietnam, is the most visible sign of devastating environmental side effects of tapioca processing (Dieu, 2003). Several kinds of wastewater from cane sugar, soft drink, milk, seafood processing, canned food factories, slaughterhouses, etc., which contain 500–8,000 mg O_2/L in terms of COD, 400–6,500 mg O_2/L as BOD_5, and a pH in the range of 4.9–5.1, contribute to the deterioration of surface water resources (these data were gathered from various processing factories in Vietnam).

Further examples show that the BOD of dairy wastes are lactose, milk fat, protein, and lactic acid, and the reported values are 0.65, 0.89, 1.03, and 0.63 kg BOD_5 per kg component, respectively (Marshall and Harper, 1984). According to Harrison et al. (1984), a broccoli and caulifl wer frozen foods manufacturer generated about 26.7 L of wastewater/kg product with BOD loading in the range of 3.3–9.3 kg/ton product (average of 5.1 kg/ton product) and total suspended solids (TSS) loading of 1.2–3.2 kg/ton product (average of 2.2 kg/ton product).

Solid Waste

Solid waste from different food processing industries is summarized in Table 2.3; canning fruits and vegetables, such as corn (660 kg/ton of product) and citrus (390 kg/ton of product), and canning of fish especially crab and shrimp (570 kg/ton of product), generates high amount of putrescible solid waste (WHO, 1993).

Beer production results in a variety of residues, such as spent grains, which have commercial value and can be sold as by-products to the agricultural sector (IFC, 2007d). Solid organic waste in dairy processing facilities mainly originates from production processes and includes nonconforming products and product losses, grid and filte residues, sludge from centrifugal separators and wastewater treatment, and packaging waste arising from incoming raw materials, and production line damage (IFC, 2007e).

Fish processing industries generate potentially large amounts of organic waste and by-products from inedible fis parts and endoskeleton shell parts from the crustacean peeling process (IFC, 2007c). Table 2.3 provides examples of solid waste loads of about 280 kg/ton of product and 570 kg/ton of product from fis and shrimp canning (WHO, 1993).

Depending on the raw materials, food and beverage processing activities may generate significan volumes of organic, putrescible solid waste in the form of inedible materials and rejected products from sorting, grading and other production processes (IFC, 2007b).

Air Emissions

Air emissions inventories from WHO (1993) indicate that total suspended particulate (TSP), combustion products, and volatile organic compounds (VOC) are the primary pollutants emitted from the food

Table 2.3. Solid waste generation from food processing industry

Process	Unit (U)	Putrescible (kg/U)	Waste composition
Slaughterhouse	ton LWK	35	Blood, paunch, hooves
	ton LWK	3	Infected animals, organs
Poultry processing	1,000 birds	35	Feathers, hooves, inedibles
Packing house	ton of product	300	Bones, inedible meat
Canning of fruits and vegetable			
Apples	ton of product	280	Peels, cores, seeds, etc.
Beets, carrots	ton of product	210	
Citrus	ton of product	390	
Corn	ton of product	660	
Olives	ton of product	140	
Peaches	ton of product	270	
Pears	ton of product	290	
Peas	ton of product	120	
Potatoes	ton of product	330	
Tomatoes	ton of product	80	
Vegetables, misc.	ton of product	220	
Canning of fis			
Fish	ton of product	280	Inedible fis parts
Crab, shrimp	ton of product	570	
Vegetable oil refinin	ton of product	4.7	Purificatio mud soaked in oil
Sugar refinerie	ton of product	N/A	Spent beats and canes
Starch and glucose	ton of product	N/A	Corn residues
Beverage industries			
Alcohol distilleries	ton of product	300	Spent resins, figs canes, etc.
Beer brewing	ton of cereal	100	Spent hop, grain, residues, yeast
	m³ of beer	20	

LWK is the live weight of animal killed.
Source: WHO (1993).

processing industry. Combustion products include nitrogen oxides (NO_X), carbon monoxide (CO), carbon dioxides (CO_2), and sulfur oxides (SO_X). Without air pollution control facilities, wheat milling, and rye milling process release high amounts of TSP, of about 38.0 kg TSP/ton, while soybean milling and corn milling also contribute about 11.73 kg TSP/ton and 6.25 kg TSP/ton, respectively. Fish processing using direct-fire driers, starch manufacturing, and beer brewing also generate about 4.0 kg TSP/ton (Table 2.4).

Potential emission sources from sugarcane processing industry include the sugar granulators, sugar conveying and packaging equipment, bulk load-out operations, boilers, granular carbon and char generation kilns, regenerated adsorbent transport systems, lime kilns and handling equipment, carbonation tanks, multieffect evaporator stations, and vacuum boiling pans (Chen and Chou, 1993).

However, there are also nuisance impacts of emissions such as noise, odor (even in cases of high acceptance rates), and dust. Although smoke and particulate matter may be problematic, odors are the most objectionable emissions from fis processing plants. The fis by-products segment results in more of these odorous contaminants than canning, because the fis are often in a further state of decomposition, which usually results in greater concentration of odors (Midwest Research Institute, 1994). The largest odor source in the fis by-products segment is the fis meal driers. As indicated in Table 2.4, direct-fire driers emit more particulate than steam tube driers. Hydrogen sulfid (H_2S) and trimethylamine [$(CH_3)_3N$] are other odorous gases emitted from fis processing.

Particulate matter (dust) and VOCs are the principal emissions from vegetable oil processing. Dust results from the processing of raw materials, including cleaning, screening, and crushing, whereas VOC emissions are caused by the use of oil extraction solvents, normally hexane (IFC, 2007a).

Waste Management

The total impacts of the waste described can be managed by aiming to (1) reduce the waste, (2) recover resources and use them, and (3) treat and discharge (Niranjan and Shilton, 1994). These waste and other reduction strategies were included in a checklist for food processors by

Table 2.4. Air emission from food processing industry

Process	Unit (U)	TSP (kg/U)	CO (kg/U)	VOC (kg/U)	NO$_X$ (kg/U)	H$_2$S (kg/U)
Meat smokehouses						
Uncontrolled	ton	0.15	0.3	0.18		58
Low voltage ESP or after burner	ton	0.15	0.0	0.075		
Fish processing (canning and manufacturing of by-products)						
Steam tube driers	ton	2.5				0.05
Direct-fire driers	ton	4.0				0.05
Grain mills						
Feed mills, uncontrolled	ton	4.9				
Wheat millings						
Uncontrolled	ton	38.0				
Cyclones and fabric filter	ton	0.8				
Drum milling, uncontrolled	ton	3.0				
Rye milling						
Uncontrolled	ton	38.0				
Cyclones and fabric filter	ton	0.8				
Oat milling, uncontrolled	ton	1.25				
Rice milling, uncontrolled	ton	2.97				
Soybean milling, uncontrolled	ton	11.73				
Dry corn milling, uncontrolled	ton	6.25				
Wet corn milling, uncontrolled	ton	6.24				
Starch manufacturing						
Uncontrolled	ton	4.0				
Controlled[a]	ton	0.01				
Alfalfa dehydrating						
Primary cyclone						
No secondary controls	ton	5				
Medium energy wet scrubber	ton	0.5				
Meal collector cyclone						
No secondary controls	ton	2.6				
Fabric filte	ton	0.03				
Pellet cooler cyclone						
No secondary controls	ton	3				
Fabric filte	ton	0.03				
Beer brewing	ton of cereal	4.0	1.3			
m^3 of beer		0.8				0.25
Wine production	m^3					0.35

If the values do not appear in the appropriate place, this often means that its value is either small or zero.

[a] TSP emission from the various corn cleaning, grinding, and screening operations can be controlled by centrifugal gas scrubber.

Source: WHO (1993).

New York State Department of Environmental Conservation Pollution Prevention Unit (2001). Benefit of appropriate waste management go beyond environmental benefits and include cost savings and resource recovery.

Means to reduce waste include good housekeeping, equipment maintenance, and design changes to improve efficien y (Niranjan and Shilton, 1994). Good housekeeping implies that managers and employees of an enterprise are diligent in ensuring that they comply with all environmental regulations. Improved management of raw material and products inventory, reduction in raw material and product loss, and provision of training to the employees can be effective means to improve housekeeping. For instance, by introducing trays to collect the dripping blood from each individual bag of frozen meat during defrosting, a meat processing company saved significant y, by efficient y collecting a resource and ensuring one of the more effluen loading (BOD) materials does not end up in treatment facilities. Approximately 30 tons of blood per year was incorporated in the production, reducing the requirement for additional water as well (Henningsson et al., 2001). In the case of beer production, precise adjustment of bottle filler or installation of a metal sheet under the filler can minimize losses of beer during fillin stage (European Commission, 1997).

A combination of equipment maintenance and design changes have been shown to help minimize waste, up to 20–30% of costs (Envirowise, 2008). Envirowise suggests that companies lose 4% of product with inefficien packaging lines alone, and overall process control on the production line can reduce costs by 5%.

Recycling, recovery, and reuse of materials are the next most preferable method to minimize waste generation. This can be done by findin a use for the materials as they are generated (usually for animal feed) or processing them to refin out the usable portions. In a study on waste minimization in palm oil industry, Vigneswaran et al. (1999) found many possibilities to recover and reuse generated wastes. These included use of palm oil waste as animal feed, recovered fibe as fuel for production of steam in boilers, sludge as fertilizer, and recovery of methane through anaerobic digestion.

Molasses, a by-product from sugar production, has many uses. The value of molasses as feed for livestock has been known almost since sugar was firs manufactured (Paturau, 1989). The work of nutritionists has indicated the feed value of molasses, especially for milking dairy cows. Increasing the proportion of molasses in the diet of milking dairy cows results in significan increases the milk yield, milk protein, and casein concentration and protein yield (Murphy, 1999). As a result, feeds based on urea, molasses, along with various combinations, have been used in India (Ranjhan et al., 1973; Daniel et al., 1984, 1986; Verma et al., 1994, 1995; Sengar et al., 1995; Dass et al., 1996a, b) and in the other countries (Elias et al., 1967; Wythes and Ernst, 1983; Gulbransen, 1985a, b). Possibilities of substitution of molasses in swine diets has also been reported by Gohl (1981), Pettersson (1992), and M'ncene et al. (1999).

Chicken eggshells generated from confectionery production are an excellent calcium source for livestock feed. In a study on mineral composition of Slovakian and Dutch chicken eggshells, Schaafsma et al. (2000) found that the major elements in chicken eggshells are calcium and magnesium. With regard to calcium content, a chicken eggshell is equal to purifie calcium carbonate, which is widely used (Kärkkäinen et al., 1997). Low levels of heavy metals (like lead and mercury) provide an advantage for chicken eggshells, because the natural calcium carbonate source may be polluted with these elements (Whithing, 1994; Faine, 1995). Thus, with a high percentage of calcium and low levels of heavy metals, chicken eggshells are a useful ingredient for calcium enrichment of livestock feed.

Fibrous residues and pulp from tapioca production can be reused in livestock feed (Dieu, 2003), alcohol, and fertilizer/compost production (Bi and Guozhen, 1996; Agu et al., 1997; Sriroth et al., 2000). Cassava roots are generally high in carbohydrate, thus are an excellent substitute of maize in feed (Khajarern et al., 1977; Oke, 1984; Cock, 1985; Sanchez, 1990).

Wasted fruits and vegetables consisting of unqualifie fruits, outer skins, etc., are all biodegradable. Therefore, reuse of such waste as raw materials (to blend with other materials) for composting (Elvira et al., 1996; Hackett et al., 1999; Rantala et al., 2000; Suh and Rousseaux, 2001; Paredes et al., 2002) or biogas production (Augenstein et al., 1994; Saha, 1994; Mæng et al., 1999; Al-Masri, 2001; Lastella et al., 2002; Bouallagui et al., 2003) is often the likeliest options.

Several researchers indicate the possibility to reuse coffee grounds as fuel (Sivetz and Desrosier, 1977; Adams and Dougan, 1985; Silva et al.,

1998), fertilizer (Kostenberg and Marchaim, 1993; Kitou and Okuno, 1999; Nogueira et al., 1999), animal feed, a polisher for painting, a carrier for insecticides and herbicides (Silva et al., 1998), raw material for activated carbon manufacture (Evans et al., 1999), and for cultivation of the mushroom *Pheurotus ostreatus* (Fan et al., 2000). However, Silva et al. (1998) stated that utilization of coffee grounds as fuel is the best option. In the 1980s coffee grounds began to be utilized as fuel in soluble coffee processing plants in Brazil (Silva et al., 1998). Coffee grounds have been considered an excellent fuel in comparison to other biomass, because they have a heat of combustion higher than others usually used. The coffee grounds with a humidity of 50% (w.b.) can be burned directly in the boiler (Pfluge, 1975) or alternatively, burned after drying to reduce the humidity to 30% (Sivetz and Desrosier, 1977; Adams and Dougan, 1985), or even to 25% (Sivetz, 1963). Approximately 18 kg of coffee grounds at 50% of humidity (w.b.) produces the same quantity of vapor as 1 gallon of fuel oil (Silva et al., 1998). Therefore, for the typical process of the production of soluble coffee, 75% of the energy required by the plant can be supplied (Pfluge, 1975). Thus, with some technological effort to improve the conditions of utilization of this fuel in the boilers, the soluble coffee industry could become self-sufficien in thermal energy (Silva et al., 1998).

Recovery and reuse of resources in waste can include additional processing steps such as separation/concentration (like centrifugation and filtration) conversion (biological or chemical), and minimization (biodegradation) (Nimi and Gimenez-Mitsotakis, 1994). These additional processing steps may add costs, energy requirements, and other inputs, so these and the benefit need to be considered and balanced.

Much of the fis waste around the world is made into fis meal after the high-value fis oil is separated out before the meal is dried. The drying is energy intensive, but gives a stable product that can be stored and moved around the world to where it is needed. The generic term for this type of process is *rendering* and is widely used with animal by-products. This requires energy for cooking but yields materials that can be re-fed to animals. The waste is cooked, the liquid is removed, usually by pressing, and then the oil and water phases are separated, often by centrifugation. The oil can be used as a source of the omega-3 fis oil. Fish oil has traditionally been used as food oil in Europe to make a number of products, including margarine.

Pectin is often recovered from fruit peels. Mangoes are one of the most popular fruits worldwide and thus peel waste is significant The peels of mangoes, like other fruit, include pectin that can be separated out and used for additional food purposes (Pedroza-Islas et al., 1994). This recovery involves filtratio of the pulp, followed by extraction of pectin, and concentration (Pedroza-Islas et al., 1994).

Membrane filtratio is commonly used with dairy whey to recover protein, lactose, and minerals (Hansen and Hwang, 2003).

Cellulosic materials are widely available waste materials (Venkatesh et al., 1994). These materials can be converted to fermentation sugars (saccharification for production of many end products including ethanol, organic acids, and edible or specialty oils (Greenwalt, 2000). Saccharificatio is typically accomplished through enzymatic, chemical, or physical means—with enzymatic conversion being the most common (Greenwalt, 2000). However, Venkatesh et al. (1994) found that simultaneous saccharificatio and fermentation improves conversion effectiveness, especially for lactic acid and other organic acid production. This is due to reduced enzymatic inhibition and development of unfavorable pH levels.

Anaerobic digesters have been used to decrease the total load of food processing waste (Banks, 1994). Such processes often produce useful products, such as methane for energy, specialty chemicals (like esters), and ingredients for food (organic acids). Anaerobic digestion is most commonly done for methane production. This is considered a three-stage process of hydrolysis, acidogenesis, and methanogenesis (Song and Hwang, 2003). These processes are carried out by symbiotic microorganisms. Methane production from this process is a high-value fuel, producing 12,000 kcal/kg with a clean burn (Song and Hwang, 2003).

By using the by-products from the food industry as fuel, one may get multiple benefits It changes a waste into an asset. It provides energy. Another example of using waste for fuel is biodiesel. Biodiesel is generally less polluting than regular diesel. Biodiesel is a specifi product that substitutes for diesel (for engines). Biofuels is the broader, generic term for fuels that come from biological materials. There are biofuels that are obtained from the relatively easy breakdown of starches and sugars, which is why most current effort focuses on materials like corn and sugarcane. But more complex carbohydrates (e.g., cellulose and lignin) also may be broken down into biofuel in the future—many of these require less energy inputs, have less negative environmental impact, and do not compete with food materials. Currently, the major end product has been ethanol, but other processes are looking at other simple fuels like butanol. Other biofuels can be obtained from lipids. After filtering fry oil can be used directly as a fuel in modifie vehicles. Biodiesel is created by modifying fatty acids and yielding glycerol as a by-product. Another approach is to use lipids such as fry oil directly as a fuel (after filtering) Direct oil burning requires a more complex engine conversion than biodiesel or ethanol-based fuels—but requires less energy and chemical inputs on an ongoing basis.

It is estimated that 50–60% of food process water can be reused (Niranjan and Shilton, 1994). The water generally needs to be treated prior to reuse. Screening, filtering or dilution with fresh water are common means to treat water for reuse (Niranjan and Shilton, 1994). Brining, heat exchanger water, flum water, and retort water are potential applications for reused process water.

In addition to reuse, water conservation is an important means to reduce environmental impacts of food processing. Niranjan and Shilton (1994) suggested processes that could be done with less or no water. Examples included mechanical conveying instead of hydraulic transport, dry cleaning rather than water, and blanching with hot gases or microwave instead of hot water (Niranjan and Shilton, 1994). Such conservation has the potential to save 40% of waste load for water-intensive operations like vegetable canning (Niranjan and Shilton, 1994).

As a result of the management options described, the fina release of any remaining waste into the environment should be minimal. Further, water discharge should be as clean as possible (low BOD and TSS) and free from any biological or nonfood components. Hansen and Hwang (2003) described the many processing options for handling remaining waste, common practices to the industry today. Land application is most common, followed by additional processing such as sedimentation, fil tration, activated sludge, and chemical treatment (Hansen and Hwang, 2003).

Energy Use

Energy use is a common concern across industries. For food processing, energy costs are much lower than raw material costs (e.g., 4.5% of total material costs for fruit and vegetable processing; Masanet et al.,

2007); however, with rising energy costs its contribution is increasing. Frito-Lay, a manufacturer of snack foods, and a division of PepsiCo, implemented a comprehensive corporate energy management program in 1999 that led to energy savings of 21% across its 34 United States facilities and saved the company more than US$40 million in energy costs (Frito-Lay, 2006). The additional considerations of the environmental impacts of energy use (e.g., use of nonrenewable resources and greenhouse gas emissions) increase the motivation to improve energy efficien y and sources of energy.

The main energy users in food processing are wet corn milling; beet sugar; soybean oil mills; malt beverages; meatpacking; canned fruits and vegetables; frozen fruits and vegetables; and bread, cake, and related goods (ICF, 2007).

Process heating and cooling systems (steam systems, ovens, furnaces, and refrigeration units) have the greatest energy requirements in food manufacturing (over 75% of the sector's energy use) (ICF, 2007). Motor-driven systems (pumps, fans, conveyors, mixers, grinders, and other process equipment) represent 12% of the sector's energy use, and facility functions (heat, ventilation, lighting, etc.) comprise approximately 8% (ICF, 2007).

As stated above, brewery processes are intensive users of both electrical and thermal energy. Thermal energy is used to raise steam in boilers, and the process of refrigeration system is typically the largest single consumer of electrical energy (IFC, 2007d). Specifi energy consumption in a brewery can vary from 300 to 342 mega joule per hectoliter (MJ/hL) depending on size, sophistication, and other factors (SSB Tool, 2008). Table 2.5 provides examples of energy consumption indicators for efficien breweries.

Table 2.5. Energy consumption of efficien breweries

Outputs per unit of product	Unit	Benchmark
Heat	MJ/hL	85–120
Electricity	kWh/hL	7.5–11.5
Total energy	MJ/hL	100–160

Input and output figure for large German breweries (capacity over 1 million hectoliter beer), EC (2006).
Source: IFC (2007d).

Table 2.6. Energy consumption of efficien dairy processing facilities

Outputs per unit of product	Unit	European dairies[a]	Swedish dairies[b]	Danish dairies[b]	Finnish dairies[b]	Norwegian dairies[b]	Industry benchmark[b]
Market milk and cultured products		0.09–1.11	0.11–0.34	0.07–0.09	0.16–0.28	0.45	0.1–0.2
Cheese and whey	kWh/L processed milk	0.06–2.08	0.15–0.34	0.12–0.18	0.27–0.82	0.21	0.2–0.3
Milk powder, cheese, and/or liquid products		0.85–6.47	0.18–0.65	0.30–0.71	0.28–0.92	0.29–0.34	0.3–0.4
Ice cream	kWh/kg ice cream	0.75–1.6	0.8–1.2				

[a] European Dairy Association (2002), cited in EC (2006).
[b] Nordic Council of Ministers (2001).
Source: IFC (2007e).

Dairy processing facilities consume considerable amounts of energy. Typically, 80% of the energy requirements are for thermal uses to generate hot water and produce steam for process application and cleaning purposes, and the remaining 20% are used as electricity to drive processing machinery, refrigeration, ventilation, and lighting (IFC, 2007e). Table 2.6 presents energy consumption data for dairy processing facilities.

Energy consumption for common fis production processes varies from 15 to 2,300 MJ/ton of raw material depending on processes. Fish meal production consumes about 2,300 MJ/ton of raw material; processing of shrimp needs 350 MJ/ton of raw material, while fille production needs only 18 MJ/ton of raw material (IFC, 2007c).

The initial cooling, processing, and cold storage of fresh fruit and vegetables are among the most energy-intensive segments of food industry. Cooling the fresh fruit and vegetables before processing removes the *field* heat from the freshly harvested products in time to inhibit decay and help maintain moisture content, sugars, vitamins, and starches, while the quick freezing of processed fresh fruit and vegetable maintain the quality, nutritional value, and physical properties for extended periods (Hackett et al., 2005). Table 2.7 provides examples of energy consumption of this sector. Depending on processes, energy consumed varies from 0.5 to 30 kWh/ton frozen vegetable.

The total impact (from all sustainability perspectives) of energy use can be reduced. This, like waste, is best done firs by reducing energy

Table 2.7.   Energy consumption of efficien food and beverage processing

Outputs per unit of product	Unit	Industry benchmark
Sorting if vegetables (carrots)		8
Caustic peeling of vegetables		2
Steam peeling of vegetables		3.5
Washing of vegetable (carrots)		2.5
Mechanical processing prior to freezing (diced carrots)	kWh/ton frozen vegetables	2.5
Drum blanching in deep freezing of vegetables		0.5–1.3
Countercurrent water cooling of vegetable		0.5–1.3
Belt blancher with water cooler		2–9
Belt blancher with air cooling		7–30

Source: IFC (2007b).

needs, then producing energy from waste, and use of renewable energy sources. The firs step to this improvement should be to assess energy usage. This becomes important when considering any changes to the operation, and ensuring the benefit–cos comparison is favorable. The Energy Star program in the United States suggests that energy efficien cies can be done at four levels:

1. At the component and equipment level, energy efficien y can be improved through regular preventative maintenance, proper loading and operation, and replacement of older components and equipment with higher efficien y models (e.g., high efficien y motors) whenever feasible.
2. At the process level, process control and optimization can be pursued to ensure that production operations are running at maximum efficien y.
3. At the facility level, the efficien y of space lighting, cooling, and heating can be improved while total facility energy inputs can be minimized through process integration and combined heat and power systems, where feasible.
4. And lastly, at the level of the organization, energy management systems can be implemented to ensure a strong corporate framework exists for energy monitoring, target setting, employee involvement, and continuous improvement (Masanet et al., 2007).

It has been estimated that, the energy savings associated with improved boiler maintenance can be up to 10%. Improved maintenance may also reduce the emissions of air pollutants (Masanet et al., 2007). The United States Department of Energy estimates that repairing leaks in an industrial steam distribution system will lead to energy savings of around 5–10% (DOE, 2006). Integrated heat recovery systems provide another means to increase energy efficien y. McCain Foods, a major producer of frozen French fried potatoes, installed an integrated heat recovery system in its Scarborough, England, facility in 1995. Heat was recovered from fryer exhaust gases via a vapor condenser and from boiler flu gases via economizers. The recovered heat was used to preheat air for potato chip dryers, to provide hot water for potato blanching, and to provide hot water for miscellaneous processes around the facility. The project led to annual energy savings of £176,000 (US$280,000 in 1995) and a simple payback period of 3.6 years (Caffal, 1995). Similar efforts across other operations in the plant can fin efficiencies described in detail by Masanet et al. (2007). For example, according to the United States Department of Energy, the typical industrial plant in the United States can reduce its electricity use by around 5–15% by improving the efficien y of its motor-driven systems (DOE, 2006).

Discussion about sources of energy from processing waste was included earlier, such as methane production for dairy waste or diesel from fats. Consideration for renewable sources of energy can further reduce the impact associated with energy usage. These sources can be generated on site or through a power provider, with solar or wind power most common. Many facilities are also taking a fina step to *neutralize* their energy impact with carbon offsets. For example, Frito-Lay purchases carbon offsets (through renewable energy credits), along with energy/emission reduction, as part of their environmental program.

Summary

Environmental impacts of food processing come primarily from waste, water use, and energy use. Waste considerations have evolved over the years from focusing on treating the waste after it has been generated to considering waste from a broader management perspective. This includes considering waste prevention and minimization at the source (no waste means no pollution and no cost involved in its management).

When waste prevention and minimization at the source can no longer be carried out, waste materials should be reused in other production processes as secondary raw materials for production of new products. Even when both approaches are firs applied, there is often still some residual waste material left. Preferably, this leftover waste has to be treated firs and the remaining has to be disposed properly to prevent environmental risks, pest problems, and endangering human health and safety. In both literature and practices from industrialized countries, one can fin that each of these approaches play an important role in environmental protection. Often the combination and integration of some or all these approaches, along with water and energy conservation, is seen as the best or only strategy to overcome continuing environmental deterioration (Dieu, 2003).

A means to bring these considerations together is a zero waste industrial ecosystem model. In this model, transfers of energy and materials exist among agricultural field and several industrial enterprises. By doing so, food processing industry and agriculture can cooperate for environmentally sound development of both (Figures 2.1 and 2.2).

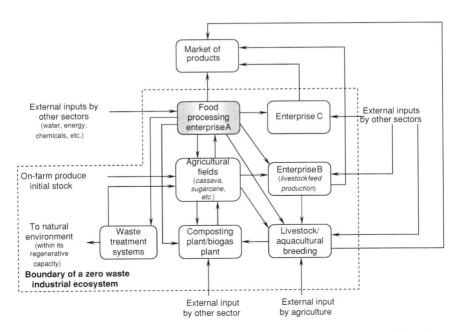

Figure 2.1.   General recommended physical model of a zero waste industrial ecosystem for food processing industry (Dieu, 2003).

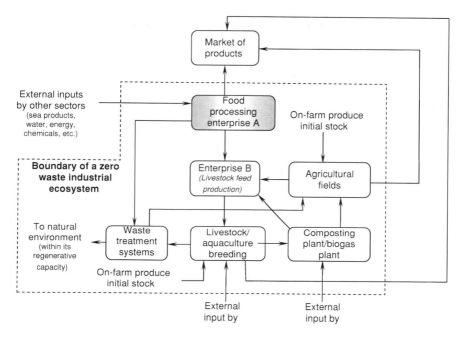

Figure 2.2. A proposed physical model of a zero waste industrial ecosystem of food processing industrial subsectors which do not use agricultural products as raw materials (Dieu, 2003).

Case Studies

Material fl w networks of the physical–technological model of a zero emission industrial ecosystem of three case studies are described in Figures 2.3–2.5. The firs case study consists of a group of tapioca-producing households in Tra Co village, Dong Nai province, Vietnam. These enterprises represent the characteristics of household (small)-scale enterprise and are all in the same industrial sector. The second case study is within the same industrial sector, a large-scale tapioca-producing plant called Tan Chau-Singapore Company in Tay Ninh province, Vietnam. These case studies help us to understand the possibilities and difficultie in developing, applying, and implementing the zero waste industrial ecosystem model in different scales of the same type of industry. The third case study focuses on a group of six different food processing enterprises located in Bien Hoa 1 industrial zone, Dong Nai province, Vietnam. It is comparable to the case study on Tra Co

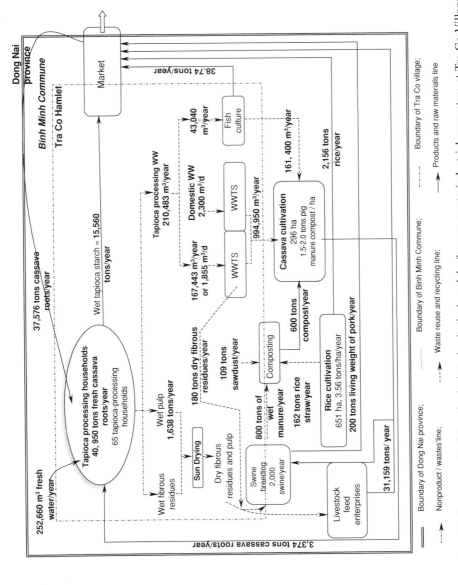

Figure 2.3. Material flow network of the physical–technological model of a zero waste industrial ecosystem at Tra Co Village (Dieu, 2003).

52

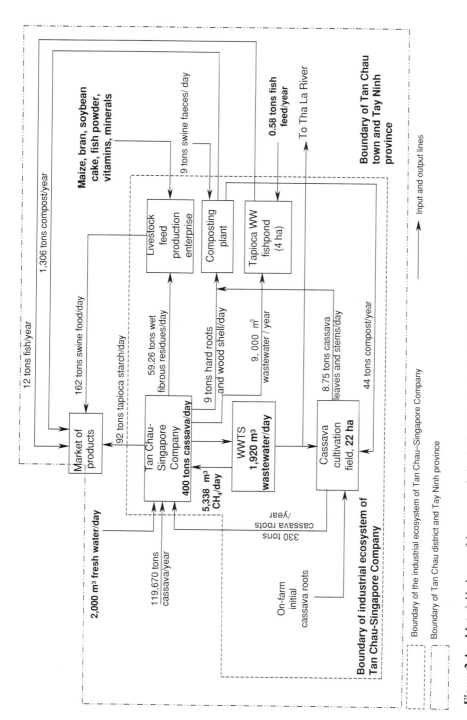

Figure 2.4. Material balance of the proposed physical–technological model of a zero emission industrial ecosystem at Tan Chau–Singapore Company (Dieu, 2003).

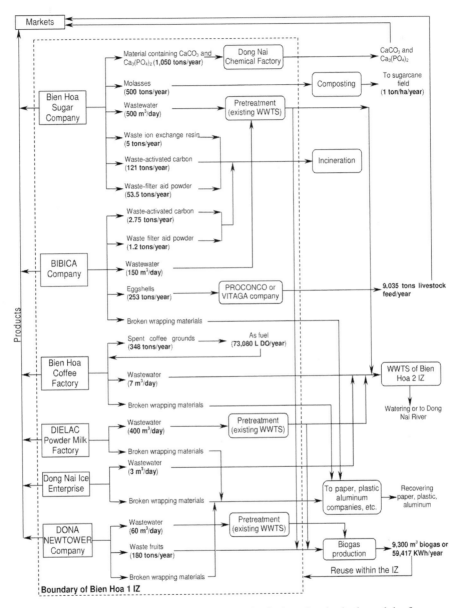

Figure 2.5. Material fl w network of the physical–technological model of a zero emission industrial ecosystem of food processing companies in Bien Hoa 1 IZ (Dieu, 2003).

village in the sense that a group of companies are studied, but differs in terms of scale of the enterprises and industrial sectors. This case study also explores the potentials of industrial zones, compared to enterprises located outside industrial zones.

In general, the scope of food processing industry extends from farms, where raw materials as crops (such as cassava, sugarcane, coffee, vegetables, and fruits) and livestock (chickens, cattle, swine) are grown, to factories, where these raw materials are processed and then brought to the markets, where the fina products are sold.

References

Adams, M.R., and J. Dougan. 1985. *Waste Products, Coffee*. London: Elsevier Applied Science Publisher.

Agu, R.C., A.E. Amadife, C.M. Ude, A. Onyia, E.O. Ogu, and M. Okafor. 1997. Technical note: Combined heat treatment and acid hydrolysis of cassava grate waste (CGW) biomass for ethanol production. *Waste Manage*. 17(1):91–96.

Al-Masri, M.R. 2001. Changes in biogas production due to different ratios of some animal and agricultural wastes. *Bioresour. Technol*. 77:97–100.

Augenstein, D., J. Benemann, and E. Hughes. 1994. *Electricity from Biogas, Six National Bioenergy Conference, 2–6 October*, Reno, NV, USA.

Banks, C.J. 1994. "Anaerobic digestion of solid and high nitrogen content fractions of slaughterhouse wastes." In: *Environmentally Friendly Food Processing*, ed. E. Gaden. New York, NY: American Institute of Chemical Engineers.

Bi, G., and Y. Guozhen. 1996. "Commercial-scale production of ethanol from cassava pulp, cassava, starch and starch derivatives." In: *Proceedings of the International Symposium held in Nanning, Guangxi, China*, Novemebr 11–15, 1996.

Bouallagui, H., R.B. Cheikh, L. Marouani, and M. Hamdi. 2003. Mesophilic biogas production from fruit and vegetable waste in a tabular digester. *Bioresour. Technol*. 86:85–89.

Caffal, C. 1995. *Energy Management in Industry*. The Netherlands: Centre for the Analysis and Dissemination of Demonstrated Energy Technologies (CADDET). Analysis Series, December 17.

Carawan, R.E., and D.H. Pilkington. 1986. *Reduction in Waste Load from a Meat Processing Plant-Beef*, pp. 27624–27695. Raleigh, NC: North Carolina State University.

Chen, J.C.P., and C. Chou. 1993. *Cane Sugar Handbook*, 12th edn. New York, NY: John Wiley & Sons, Inc.

Clanton, C.J., P.R. Goodrich, B.D. Backus, E.J. Fox, and H.D. Morris. 1994. "Anaerobic digestion of cheese whey." In: *Environmentally Responsible Food Processing*, ed. E. Gaden. New York, NY: American Institute of Chemical Engineers.

Cock, J.H. 1985. *Cassava New Potential for a Neglected Crop*. Boulder and London: Westview Press.

Daniel, S.J., Q.Z. Hasan, K. Nath. 1984. Urea molasses liquid diet as a feed for calves. *Indian J. Anim. Sci.* 50:149–151.

Daniel, S.J., Q.Z. Hasan, and K. Nath.1986. Compensatory growth in crossbred calves fed on urea molasses liquid diet. *Indian J. Anim. Sci.* 56:979–981.

Dass, R.S., A.K. Verma, and U.R. Mehra. 1996a. Compensatory growth and nutrient utilization in crossbred heifers during revival period fed urea molasses liquid diet. *Asian-australas. J. Anim. Sci.* 9:563–566.

Dass, R.S., A.K. Verma, and U.R. Mehra. 1996b. Effect of feeding urea molasses liquid diet on nutrient utilization, rumen fermentation pattern and blood profil in adult male buffaloes. *Buffalo J.* 12:11–22.

Dieu, T.T.M. 2003. *Greening Food Processing Industry in Vietnam*, Ph.D. Dissertation, Wageningen University, The Netherlands.

DOE (U.S. Department of Energy). 2006. *Save Energy Now in Your Steam Systems.* Washington, D.C.: Offic of Energy Efficien y and Renewable Energy, Industrial Technologies Program. Report DOE/GO-102006-2275.

Elias, A., T.R. Preston, M.B. Willis, and T.M. Sutherland. 1967. By-products of sugar-cane and intensive beef production. 4. Fatting bulls with molasses and urea in place of grain in diets with little fibre *Rev. Cubana Sci. Agric.* 5:59.

Elvira, C., L. Sampedro, J. Dominguez, and S. Mato. 1996. Vermi-composting of wastewater sludge from paper-pulp industry with nitrogen rich materials. *Soil Biol. Biochem.* 29(3–4):759–792.

Envirowise. 2008. *Food and Drink Essentials: Environmental Information for the Industry.* UK: Envirowise.

Erkman, S. 1997. Industrial ecology: An historical view. *J. Clean. Prod.* 5(1–2):1–10.

European Commission. 1997. *Final Report on Cleaner Technologies for Waste Minimization.* Luxembourg: European Commission.

European Commission. 2005. *Integrated Pollution Prevention and Control, Reference document on Best Available Techniques in the Slaughterhouses and Animal byproduct Industries.* BREF, European Commission, May, 2005. Available from http://www.ifc.org/ifcext/sustainability.nsf/AttachmentsByTitle/gui_EHSGuidelin es2007_PoultryProcessing/$FILE/Final+-+Poultry+Processing.pdf. Accessed Sept ember 15, 2008.

Evans, M.J.B., E. Haliop, and J.A.F. MacDonald. 1999. The production of chemically activated carbon.*Carbon* 37:269–274.

Faine, M.P. 1995. Dietay factors related to preservation of oral and skeletal bone mass in women. *J. Prosthet. Dent.* 73:65–72.

Fan, L., A. Pandey, R. Mohan, and C.R. Soccol. 2000. Use of various coffee industry residues for the cultivation of pleutotus ostreatus in solid state fermentation. *Acta Biotechnol.* 20(1):41–52.

Frijns, J., P.T. Phuong, and A.P.J. Mol. 2000. *Ecological Modernization Theory and Industrializing Economies: The Case of Vietnam, Ecological Modernization Around the World: Perspectives and Critical Debates*, pp. 257–292. London, Portland: Frank Cass.

Frito-Lay. 2006. *Frito-Lay Recognized by U.S. EPA and U.S. DOE.* Press Release. Plano, Texas. June 5, 2006.

Fryer, P. 1995. *Clean Technology in the Food Industry, Clean Technology and the Environment*, pp. 254–276. London: Blackie Academic & Professional.

Gohl, B. 1981. *Tropical Feeds*. FAO (Food and Agriculture Organization) of the United Nations.

Greenwalt, C. 2000. *Dissertation: Utilization of Crop Residue and Production of Edible Single cell Oil for an Advanced Life Support System*. Ithaca, NY: Cornell University.

Gulbransen, B. 1985a. Survival feeding of cattle with molasses. 1. Feeding of non-pregnant heifers with molasses plus urea and roughage. *Aust. J. Expt. Agric.* 25:1–3.

Gulbransen, B. 1985b. Survival feeding of cattle with molasses. 2. Feeding steers with molasses/urea plus either sorghum grain (*Sorghum vulgare*) or cotton seed meal (*Gossipum hirsutum*). *Aust. J. Expt. Agric.* 25:4–8.

Hackett, B., S. Chow, and A.R. Ganji. 2005. *Energy Efficiency Opportunities in Fresh Fruit and Vegetable Processing/Cold Storage Facilities.* Paper presented in the 2005 ACEEE Summer Study on Energy Efficien y in Industry, July, 19–22, 2005, West Point, NY.

Hackett, G.A.R., C.A. Easton, and S.J.B. Duff. 1999. Composting of pulp and paper mill fl ash with wastewater treatment sludge. *Bioresour. Technol.* 70:217–224.

Hansen, C.L., and S. Hwang. 2003. *Waste Treatment. Environmentally-Friendly Food Processing*. Cambridge, England: Woodhead Publishing Limited.

Harrison, J.R., L.A. Licht, and R.R. Peterson. 1984. "The treatment of waste from the fruit and vegetable processing industries." In: *Surveys in Industrial Wastewater Treatment: Food and Allied Industries*, eds D. Barnes, C.F. Forster, and S.E. Hrudey. Pitnam Boston, London, Melbourne: Advanced Publishing Program.

Henningsson, S., A. Smith, and K. Hyde. 2001. Minimizing material fl ws and utility use to increase probability in the food and drink industry. *Trends Food Sci. Technol.* 12:75–82.

Higgins, T.E. 1995. *Pollution Prevention Handbook*. Boca Raton, FL: Lewis Publishers.

Hrudey, S.E. 1984. "The management of wastewater from the meat and poultry products industry." In: *Surveys in Industrial Wastewater Treatment: Food and Allied Industries*, eds D. Barnes, C.F. Forster, and S.E. Hrudey. USA: John Wiley & Sons Inc.

ICF. 2007. *Energy Trends in Selected Manufacturing Sectors: Opportunities and Challenges for Environmentally Preferable Energy Outcomes.* Available from http://www.epa.gov/sectors/pdf/energy/ch3-4.pdf. Accessed February 2008.

International Finance Corporation (IFC). 2007a. *Environmental, Health and Safety Guidelines: Vegetable Oil Processing*. World Bank Group.

International Finance Corporation (IFC). 2007b. *Environmental, Health and Safety Guidelines: Food and Beverage Processing*. World Bank Group.

International Finance Corporation (IFC). 2007c. *Environmental, Health and Safety Guidelines: Fish Processing*. World Bank Group.

International Finance Corporation (IFC). 2007d. *Environmental, Health and Safety Guidelines: Breweries*. World Bank Group.

International Finance Corporation (IFC). 2007e. *Environmental, Health and Safety Guidelines: Dairy Processing*. World Bank Group.

Kärkkäinen, M.U.M., J.W. Wiersma, and C.J.E. Lamberg-Allardt. 1997. Postprandial parathyroid hormone responses to four calcium-rich foodstuff. *Am. J. Chin. Nutr.* 65:1726–1730.

Khajarern, S., J.M. Khajarern, and Z.O. Muller. 1977. "Cassava in the nutrition of swine." In: *Proceedings of a Workshop on Cassava as Animal Feed, held as the University of Guelph, 18–20 April 1977.*

Kitou, M., and S. Okuno. 1999. Decomposition of coffee residue in soil. *Soil Sci. Plant Nutr.* 45(4):981–985.

Kostenberg, D., and U. Marchaim. 1993. Anaerobic digestion and horticultural value of solid waste from manufacture of instant coffee. *Environ. Technol.* 14:973–980.

Lastella, G., C. Test, G. Cornacchia, M. Notornicola, F. Voltasio, and V.K. Sharma. 2002. Anaerobic digestion of semi-solid organic waste: Biogas production and its purification *Energy Convers. Manage.* 43:63–75.

Mæng, H., H. Lund, and F. Hvelplund. 1999. Biogas plants in Denmark: Technological and economic developments. *Appl. Energy* 64:195–206.

Mai, H.N.P., L.N. Thai, N.T. Viet, and G. Lettinga. 2001. *Effect of Organic Loading Rate on Treatment Efficiency for Tapioca Processing Wastewater Using UASB, International Conference: Industry and Environment in Vietnam, April 20-21, 2001,* Ho Chi Minh City, Vietnam.

Marshall, K.R., and W.J. Harper. 1984. "The treatment of wastes from the dairy industry." In: *Surveys in Industrial Wastewater Treatment: Food and Allied Industries,* eds D. Barnes, C.F. Forster, and S.E. Hrudey. Boston, London, Melbourne: Pitnam Advanced Publishing Program.

Masanet, E., E. Worrell, W. Graus, and C. Galitsky. 2007. *Energy Efficiency Improvement and Cost Saving Opportunities for the Fruit and Vegetable Processing Industry: An ENERGY STAR® Guide for Energy and Plant Managers.* Available from http://www.energystar.gov/index.cfm?c=in_focus.bus_food_proc_focus. Accessed September 15, 2008.

Midwest Research Institute. 1994. *Final Report of Emission Factor Documentation for AP-42.* Section 9.13.1, Fish Processing, EPA Contract No. 68-D2-0159, Work Assignment No. I-08, MRI Project No. 4601-08, March 1994.

M'ncene, W.B., J.K. Tuitoek, and H.K. Muiruri 1999. Nitrogen utilisation and performance of pigs given diets containing a dried or undried fermented blood/molasses mixture. *Anim. Feed Sci. Technol.* 78:239–247.

Murphy, J.J. 1999. The effects of increasing the proportion of molasses in the diet of milk dairy cows on milk production and composition. *Anim. Feed Sci. Technol.* 78:189–198.

New York State Department of Environmental Conservation Pollution Prevention Unit. 2001. *Environmental Self-Assessment for the Food Processing Industry: A Quick and Easy Checklist of Pollution Prevention Measures for the Food Processing Industry.* Available from http://www.dec.ny.gov/docs/permits_ej_operations_pdf/esafood.pdf. Accessed September 15, 2008.

Nimi, D., and P. Giminez-Mitsotakis. 1994. "Creative solutions for bakery waste effluent." In: *Environmentally Friendly Food Processing,* ed. E. Gaden. New York, NY: American Institute of Chemical Engineers.

Niranjan, K., and N.C. Shilton. 1994. "Food processing wastes—their characteristics and an assessment of processing options." In: *Environmentally Responsible Food Processing*, ed. E.L. Gaden. New York, NY: American Institute of Chemical Engineers.

Nogueira, W.A., F.N. Nogueira, and D.C. Devents. 1999. Temperature and pH control in composting of coffee and agricultural wastes. *Water Sci. Technol.* 40(1): 113–119.

Nordic Council of Ministers. 2001. *Best Available Techniques (BAT) for the Nordic Dairy Industry*. Copenhagen: Nordic Council of Ministers.

Oanh, L.T.K., K. de Jong, H.N.P. Mai, and N.T. Viet 2001. Removing suspended solids from tapioca processing wastewater in upfl w anaerobic filte (UAF). *International Conference: Industry and Environment in Vietnam*, Ho Chi Minh City, Vietnam.

Oke, O.L. 1984. The use of cassava as pig feed. *Nutr. Abstr. Rev.—Series B* 54(7):301–314.

Paredes, C., M.P. Bernal, J. Cegarra, and A. Roig. 2002. Bio-degradation of olive mill wastewater sludge by its co-composting with agricultural wastes. *Bioresour. Technol.* 85:1–8.

Paturau, J.M. 1989. *By-Products of the Cane Sugar Industry: An Introduction to Their Industrial Utilization*, 3rd Completely Revised Edition. Amsterdam-Oxford-New York-Tokyo.

Pedroza-Islas, R., E. Aguilar-Esperanza, and E.J. Vernon-Carter. 1994. "Obtaining pectins from solids waste derived from mango (*Mangifera indica*) processing." In: *Environmentally Friendly Food Processing*, ed. E. Gaden. New York, NY: American Institute of Chemical Engineers.

Pettersson, A. 1992. *The Use of Sugar Cane Molasses in Diets for Pigs*. Swedish University of Agricultural Sciences, International Rural Development Centre.

Pfluge , R.A. 1975. *Solid Waste: Origin, Collection, Processing and Disposal*, pp. 365–376. Toronto: Wiley Intersciene Publishers.

Ranjhan, S.K., P.C. Sawhwey, and M.M. Jayal. 1973. *Application of Live Saving Research in Animal Feeding*. New Delhi, India: Farm Information Unit, Directorate of Extension, Ministry of Agriculture.

Rantala, P.R., K. Vaajasaari, R. Juvonen, E. Schultz, A. Joutti, and R. Mäkelä-Kurtto. 2000. Composting of forest industry wastewater sludge for agricultural use. *Water Sci. Technol.* 40(11–12):187–194.

Saha, S.L. 1994. Promoting use of biogas in India. *Electrical India* 34:13–16.

Sanchez, S. 1990. *The Use of Cassava as Animal Feed in Developing Countries— Implications on Food Security and Balance of Payments*. Report for the Food and Agriculture Organization of the United Nations, Rome, Italy.

Schaafsma, A., I. Pakan, G.J.H. Hofstede, F.A.J. Muskiet, E. Van Der Veer, and P.J.F. DeVries. 2000. Processing and products: Mineral, amino acid and hormonal composition of chicken eggshell powder and the evaluation of its use in human nutrition. *Poult. Sci.* 79:1833–1838.

Sengar, S.S., A.K. Verma, V.P. Varhney, and U.R. Mehra. 1995. Effect of feeding urea molasses liquid diet feeding on the growth performance, blood profil and thyroid gland activity in buffalo heifers in early stage of growth. *Buffalo J.* 11:157–163.

Silva, M.A., S.A. Nebra, M.J. Machado Silva, and C.G. Sanchez. 1998. The use of biomass residues in the Brazilian soluble coffee industry. *Biomass Bioenergy* 14(5/6):457–467.

Sivetz, M. 1963. *Coffee Processing Technology*, Vol. 2. Wesport, CT: AVI Publishing Co.

Sivetz, M., and N.W. Desrosier. 1977. *Coffee Technology*. Wesport, CT: AVI Publishing Co.

Song, M., and S. Hwang. 2003. "Recycling food processing wastes." In: *Environmentally-friendly Food Processing*, eds B. Mattsson and U. Sonesson. Cambridge, England: Woodhead Publishing Limited.

Sriroth, K., R. Chollakup, S. Chotineeranat, K. Piyachomkwan, and C.G. Oates. 2000. Processing of cassava waste for improved biomass utilization. *Bioresour. Technol.* 71:63–69.

SSB Tool. 2008. *Sample of Consumption Data— Breweries*. Available from http://www.brewingschool.dk/uploads. Accessed January 31, 2008.

Suh, Y.J., and P. Rousseaux. 2001. An LCA of alternative wastewater sludge treatment scenarios, resources, conservation and recycling. *Resour. Conserv. Recy.* 35:191–200.

Venkatesh, K.V., P.C. Wankat, and M.R. Okos. 1994. "Kinetic model for lactic acid production form cellulose by simultaneous fermentation and saccharification" In: *Environmentally Friendly Food Processing*, ed. E. Gaden. New York, NY: American Institute of Chemical Engineers.

Verma, A.K., R.S. Dass, and U.R. Mehra. 1994. Response of urea molasses liquid diet feeding on growth performance, nutrient utilization and thyroid gland activity in crossbred heifers. *World Rev. Anim. Prod.* 29:101–107.

Verma, A.K., U.R. Mehra, R.S. Dass, V.P. Varhney, and H. Kumar. 1995. Performance of crossbred heifers suring revival period after long term scarcity feeding. *J. Appl. Anim. Res.* 8:63–70.

Vigneswaran, S., V. Jegatheesan, and C. Visvanathan. 1999. Industrial waste minimization initiatives in Thailand: Concepts, examples and pilot scale trials. *J. Clean. Prod.* 7:43–47.

Whithing, S.J. 1994. Safety of some calcium supplements questioned. *Nutr. Rev.* 52:95–105.

World Bank. 1998. *Breweries, Pollution Prevention Abatement Handbook*. Effective July, 1998. Available from http://www.gcgf.org/ifcext/enviro.nsf/Attachments ByTitle/gui_breweries_WB/$FILE/breweries_PPAH.pdf. Accessed September 15, 2008.

World Health Organization (WHO). 1993. *Assessment of Sources of Air, Water, and Land Pollution: A Guide to Rapid Source Inventory Techniques and Their Use in Formulating Environmental Control Strategies, Part One: Rapid Inventory Techniques in Environmental Pollution*. Alexander P. Economopoulos, Democritos University of Thrace.

Wythes, J.R., and A.J. Ernst. 1983. Molasses as a drought feed. *Proc. Aust. Soc. Anim. Prod.* 15:213–276.

Chapter 3

Distribution

Rich Pirog, Ben Champion, Tim Crosby, Sara Kaplan, and Rebecca Rasmussen

Introduction

Distribution is fundamental to modern food systems. It binds the spaces and locales of the food system together. Food distribution technologies and logics also shape the economic, social, resource, and food quality conditions of food systems. This chapter is dedicated to exploring contemporary issues of sustainability in food distribution systems. In order to do so, we must consider both the (1) direct resource, social, and metabolic impacts of food distribution, and (2) the indirect impacts of food distribution in other parts of the food chain (the viability of organic vs conventional agriculture, for instance). In addition, we must balance attention to economic, social, and environmental conditions when considering issues of sustainability.

Our efforts in this chapter primarily focus on the direct impacts of food distribution. In particular, sustainability indicators are an interesting way of framing and accounting for the direct impacts of food distribution. MacRae et al. (1989) have proposed an interesting three-tier classification system for sustainability indicators, summarized as follows:

- First-tier indicators focus largely on minimizing the impacts of existing methods, activities, or processes (efficiency improvements).
- Second-tier indicators measure the extent to which older methods, technologies, or processes with high negative impacts are being replaced by newer ones with less negative impact (substitution).

61

- Third-tier indicators help to measure the extent to which the rules and logics of industries are being rewritten with sustainability as their foundation (redesign).

In terms of transportation fuels, an example of a first-tier change indicator would be reductions in fuel used in distribution due to shorter routes or more fuel-efficient transport vehicles. A second-tier example would be the substitution of a new renewable fuel (such as biodiesel) to power vehicles to distribute food. The third-tier change might be moving from global and national to regional food distribution systems that combine self-sufficiency with dependency on regional infrastructure to provide the majority of foodstuffs. Throughout this chapter, we consider the relevance of sustainability indicators to the key challenges of sustainability in distribution that we have observed.

Before doing so, it would of course be helpful to have a working definition of food distribution and a discussion of the various elements that constitute this definition. For the purposes of this chapter, we have defined food distribution as the collective transportation and/or storage methods that join agricultural production with downstream supply chain processes such as food processing/manufacturing, retailing, and consumption. This includes raw food ingredients as well as final food products and everything in between. It is impossible to comprehensively describe all food distribution methods under this definition within a chapter of this length, much less evaluate their sustainability implications. We have necessarily simplified a complex picture, but we will present what we think are the core tensions at the heart of sustainability in the food distribution sector.

As an indication of the complexity of this picture, consider the various modes of transport and storage available to contemporary food distribution. Rail and sea transport are commonly employed for long-distance bulk transportation of relatively stable food commodities, while transportation by road and air can be faster and/or more flexible than by rail or sea. Proper storage and handling are vital for all food products, which by their very nature tend to spoil or degrade in uncontrolled conditions. Much effort and capital is therefore expended in managing storage conditions in warehouses and shipping vessels, particularly in the areas of refrigeration and modified atmospheres. Refrigeration slows the metabolic rates that lead to spoilage and bacterial growth in commodities that are in essence alive, while modified/controlled atmospheres delay spoilage and degradation. However, there are limits to the control that these technologies offer. As technology and supply chain management push these limits, technological and management sophistication (e.g., ISO 9000 standards for international trade) are required for maintaining quality and reliability.

Transportation and storage methods also serve particular development agendas within food systems. The use of transportation and storage technologies bridges the places and times of food production and consumption. In doing so, they conserve and translate the value of foods and enable the exchange of value between specific places/times. This is the economic function of food distribution, enabling the distribution of value and resources. Whether it is a local truck farmer selling at a farmers' market or a massive corporate feedlot participating in a transnational conglomerated grain–livestock joint venture system (Hendrickson and Heffernan, 2002), these principles hold. Food distribution should be viewed as setting the conditions and logistical foundation supporting major developments in food systems (e.g., the grain/feed/livestock complex and others) in advanced contemporary food systems. In the same way, food distribution is also critical for emerging systems of production and consumption such as organic agriculture, local foods, and more.

Such development agendas vary from commodity to commodity, largely depending on the biological nature of the commodity and the geographies of production, processing, and consumption for that commodity. Many fresh fruits and vegetables, grains, dairy, oils and fats, frozen foods, and fresh and frozen meats have distinctive regional geographies for production and consumption, as well as distinctive biophysical and regulatory requirements for maintaining quality. Such geographical, biophysical, and economic diversity results in an astounding complexity of distribution functions and challenges that our food system negotiates on a continual basis. Our interests in environmental sustainability must be overlaid on top of this complex backdrop.

Can the existing operational functions of food distribution systems be reengineered in environmentally sustainable ways? Furthermore, can we develop distribution logics that make it easier, or even fundamentally necessary, for the various productive and consumptive sectors of the food system to reorganize in sustainable modes of operation? To even begin to answer these questions, we must first consider what it means to be sustainable. The following section outlines the sustainability indicators we will use as benchmarks toward sustainability before moving on to apply them in exploring the gap between current distribution systems and those of a more sustainable food system.

Sustainability is a concept that has become increasingly incorporated into corporate performance measures (Gerbens-Leenes et al., 2003). However, sustainability is often a vague, ill-defined, and contested term, especially in that measurements of sustainability are not widely agreed upon. The literature has cited various types of indicators, as presented in Table 3.1, that have been used to measure the sustainability of industries or various enterprises related to the activities of the food distribution sector. These indicators have been grouped according to their main focus within economic, environmental, or social impact. Environmental sustainability indicators were most often cited as measurement tools. This may be attributed to the increased use of life cycle analysis to assess environmental impacts of different processes.

Within the realm of environmental indicators, food miles, energy consumption, and carbon dioxide emissions are the three most frequently mentioned indicators related to distribution. Food miles refer to the distance that food travels to get from place to place. For instance, on average fresh produce traveling from the continental United States into the Upper Midwest will travel 1,500 miles (Pirog et al., 2001; Pirog and Benjamin, 2003). This indicator not only shows how far food travels, but can also be used (with caution) as a crude proxy for transportation costs and emissions, assuming comparisons are made between identical modes of transport. Energy consumption includes both the fuels necessary for transportation and the energy required for the various storage conditions of the food. Carbon dioxide emissions are incorporated into these aspects as well.

Among the economic indicators, profitability, average wages, employment, and the percentage of food lost to mishandling were the most frequently cited indicators. These indicators might seem intuitive to the reader. An unprofitable operation is not likely to be sustainable. A high percentage of food lost would reflect problems within the system, and decreased profits. These indicators would seem to be food industry measurements that already exist.

The social indicators most frequently cited included the nutritional value of food being distributed and local autonomy or employment with the distribution sector. Nutritional value of food can refer to the ratio of processed foods with a longer shelf life to the distribution of fresh, perishable food. Local autonomy or employment is a proxy for how distribution impacts the local economy.

Table 3.1. Sustainability indicators

Sustainability indicators	Heller and Keolian	Pretty and others	Jones	Gerbens-Leenes	Hinrichs	Ilbery	Bellows	Jackson	Maxime	Halsberg	Clarke and others	Mintcheva	Yakovleva and Flynn	Monteiro	Cowell and Parkinson	Pirog and others
Environmental																
Food miles	X	X						X		X						X
Energy consumption			X	X				X	X				X			
CO$_2$ emissions			X	X					X							
Land usage				X											X	X
Eco-efficiency									X						X	X
Waste generated													X			
Environmental monitoring systems												X				
Hazardous substance exposure																
Health and safety incidents													X	X		
Environmental reporting													X			
Ethical transport														X		
Economic																
Waste produced per unit food	X															
Percent of food lost to mishandling	X															
Types of distribution (infrastructure)													X			
Retail access										X	X					
Revenue per square foot											X					
Food lost												X				

(Continued)

65

Table 3.1. *(Continued)*

Sustainability indicators	Heller and Keolian	Pretty and others	Jones	Gerbens-Leenes	Hinrichs	Ilbery	Bellows	Jackson	Maxime	Halsberg	Clarke and others	Mintcheva	Yakovleva and Flynn	Monteiro	Cowell and Parkinson	Pirog and others
Output growth	X												X			
Labor productivity	X												X			
Diversity and structure of market	X												X			
Imported vs domestic products	X												X			
Profitability	X												X	X		
Distribution of imports by country																
Employment													X	X		
Average wages													X	X		
Transport efficiency												X				
Intensity												X				
Social																
Distance between grower and distributor	X					X										
Profits between farmer/processor/retailer	X															
Quality of life and worker satisfaction	X															
Nutritional value of food	X					X								X		
Food safety					X									X		
Number of farmers' markets						X	X									
Local autonomy/employment						X								X		
Fair trade initiatives													X			

Many sustainability efforts include considerations that cross these three sets of indicators. For instance, food lost to mishandling or spoilage is not only an economic loss but can also include the environmental impact of methane gas emissions from decomposition in a landfill, and the social impact of a loss of food that can be rescued to service hunger and poverty programs, or, if it is degraded, sent to municipal composting programs for an environmental benefit.

Within the literature, the use of social indicators is less developed than economic or environmental indicators. Perhaps it is due to their more qualitative nature. Food miles can be quantified, as can profitability. Many of the social indicators will not return with a number. However, relative comparisons can be made of indicators across different systems. By comparing systems across these indicators, one can get relative measurements for improvements in the economic, environmental, and social spheres of sustainability.

The remainder of this chapter divides food distribution into three sections according to three scales of activity (international, national, and regional/local) in order to present the current practices and issues of sustainability at each of the scales of activity. Each section will present a general description of the typical distribution systems and organization of that scale, as well as some key challenges to sustainability involved in these distribution systems. In addition, we present case studies from each scale based on interviews we have conducted as a way of highlighting current challenges and sustainability indicators in the field. The case studies reflect many of common indicators emphasized by the academic work above, including reducing carbon footprints and greenhouse gas emissions, decreasing solid waste, and reductions in fuel and electricity, improved efficiencies for cost savings as economic measures, and even some social indicators in half of the case studies. However, most sustainability efforts in food distribution appear to focus on first-tier approaches as noted above, leaving many of the profound challenges of developing distribution alternatives and reorganizing our food systems relatively untouched.

International Distribution

Visiting a grocery store in any part of the industrialized world illustrates the results of global distribution of food. For instance, a supermarket in wintry Iowa sells bananas from Guatemala and tomatoes from Mexico. A supermarket in Ecuador contains apples from Washington state and grapes from Chile. Global food sales reached $4.1 trillion in 2002, accounting for retail sales and food service sales. Among the retail sales, fresh food accounted for US$531 billion, while processed food, including beverages and packaged foods, totaled US$1.7 trillion. Supermarkets accounted for the largest share of this amount, capturing roughly 30% of the retail market (Regmi and Gehlhar, 2005). The rise of supermarkets, shipping and storage technology innovation, the growth of urban areas, and international trade—all play a large role in global food distribution.

Grocery stores are an ever-increasing part of the global food distribution system, growing most prominently in Eastern Europe, Asia, and Latin America (Reardon et al., 2003). The growth of supermarkets has led significantly to centralized distribution systems. As the number of stores in a supermarket chain increases, there is a tendency to shift from an individual store procurement system to a centralized distribution system serving a region, such as a city, state, country, or multinational area. This centralization leads to the use of central warehouses for distribution to individual stores. Centralization can increase efficiency and decrease overall product costs by reducing coordination and other transaction costs, although it may also increase transport costs and energy usage from extra movement of the product (Regmi, 2001). Centralized distribution systems are also encouraged by the growth in both government and private standards for food quality and safety. Such standards are designed to prevent widespread food-borne illnesses and contamination. These standards act to facilitate distribution by standardizing product requirements, and are used to coordinate along supply chains (Henson and Reardon, 2005). Centralization helps reduce the costs associated with meeting the standards, and also helps coordinate logistics along the supply chain.

Centralized distribution is not the only way that the growth of supermarkets has affected the distribution sector. Supermarkets increasingly are seeking out alliances with suppliers and wholesalers in order to decrease transaction costs. Reardon et al. (2005) stated that the growing consolidation of supermarkets, combined with the goal of supermarkets to have year-round supplies of various fresh products, has led to horizontal joint ventures and other strategic alliances between firms in the northern and southern hemispheres. These supplier alliances that form partnerships between the different hemispheres are no longer constrained in meeting the necessary volume and year-round demand of the retail sector. For example, these alliances help facilitate year-round availability of tropical fruit. Mangoes available for purchase in the United States come from a variety of Central and South American countries. The mango season in Ecuador is usually October through January, while in Mexico the season lasts from March through September. Alliances between firms in the two continents help provide for year-round mango sales in the United States.

Technology innovation has also spurred this growth in global food distribution. Shipping containers have been a major factor in the reduction of world transportation costs since 1950s. Refrigerated mobile containers known as *reefers* were developed in the 1960s. These containers enabled ocean shipments of cooled and frozen cargo, namely, perishable items such as meats and produce. The rise in air cargo also sped up the growth in transport of perishable items. For example, the United States export of perishable products increased from US$3.5 billion in the fiscal year in 1989 to US$10.3 billion in the fiscal year in 1999, while perishable imports totaled US$13.1 billion in 1999, with horticultural products (including fresh vegetables, fruit and juice, bananas, cut flowers, and nursery stock) accounting for about 60% (Regmi, 2001). In addition, the increasing availability and decreasing costs of satellite technologies such as global positioning systems allow shippers to electronically track their cargo around the world. These technologies and others allow for shippers and carriers to monitor quality of product, reduce costs, and shorten delivery time (Regmi, 2001).

Global distribution has also been affected by increasing urbanization. The growth of urban populations around the globe, the rise of women in the workplace, and rising incomes are some of the lifestyle changes that have impacted global distribution (Reardon et al., 2003). These trends have all significantly contributed to the growth in supermarkets, a major player in global distribution. Urban populations have increased globally, rising most rapidly in developing nations. These populations tend to eat higher-value foods such as meats and vegetables, rather than the high cereal-based diet of rural populations. The rise of women in the workplace has also led to changes in dietary and consumption habits (Regmi, 2001). Rising incomes have also affected the nature of items distributed. Increasing growth in ownership of refrigerators and microwaves has led to the increased prevalence of packaged and processed products in developing nations (Regmi and Gehlhar, 2005).

Infrastructure

The infrastructure that supports global food distribution varies among different regions of the world. Developed countries, in general, have well-developed public infrastructure such as roads, railways, and airports. The next section on national distribution highlights the United States transportation infrastructure as an example. Private infrastructure, such as packing houses and shipping equipment linking suppliers and distributors, may be adequate as well. However, in developing countries, infrastructure can provide significant constraints arising from inadequate public infrastructure. Private infrastructure may also be inadequate. Infrastructure for meeting quality standards, such as laboratories, and certifying agents may also be lacking. Often, private infrastructure, including trucks, barges, and packing plants in developing nations, may be foreign owned (Reardon et al., 2003).

Policy and Trade

Global distribution is encouraged by the current political atmosphere promoting international free trade (Henson and Reardon, 2005). However, differing food safety regulations between nations can pose a barrier to global distribution. Tariffs also pose a barrier to global distribution. Trade in processed products is often more restrictive than in raw commodities. Tariffs are usually more expensive on processed products than raw ingredients (Regmi and Gehlhar, 2005).

Sustainability

Supermarkets, technology innovation, infrastructure, urbanization, and trade—all have significant impacts on the global food distribution system. These influencing factors pose both challenges and opportunities to those looking at creating a sustainable global distribution system. Supermarkets have implemented standards to protect food safety and quality. Technology innovation has enabled the dispersion of products worldwide. However, worldwide shipping can produce large amounts of greenhouse gases and energy usage. With the increased use of carbon footprinting and detailing of energy consumptions for food products, indicators may be the best way to judge global food systems. Schlich and Fleissner (2005) found that imported lamb from New Zealand had a lower energy turnover than regionally produced lamb in Germany, even accounting for the shipping from New Zealand. While sustainable distribution systems are essential in creating sustainable food systems, it is important to remember that distribution is but one part of a sustainable global food system.

There are more aspects to sustainability than energy usage, greenhouse gases, and food miles, though. The following case studies will look at how sustainability is being addressed in a company that has global operations.

Case Study: Adina For Life, Inc. (http://www.adinaworld.com)

Background

Launched in 2005, Adina For Life is the outgrowth of Greg Steltenpohl, founder of Odwalla, and Magatte Wade-Marchand, a native of Senegal. Today, Adina is a private company that produces and distributes fair trade and organic world tea and juice beverages to consumer markets. The company's mission is to brand and introduce fair trade practices into mainstream consumer packaged goods. In 2006, the company acquired a San Francisco Bay Area beverage distribution company to support localized brand distribution.

Adina's motto is "Drink No Evil" which the company follows by working with fair trade sources and small producers around the world. Adina's initial motivation, to help traditional producers produce traditional products, makes it one of only a few companies that started down the sustainability path on the social, compared to the environmental, leg; a leg on which many find a much harder to make significant progress.

Adina and Sustainability

Adina has four main organizational goals that are deeply intertwined with sustainability:

1. Work with community-based cooperatives and collectives.
   - *Put the power to enrich and protect their community into the hands of the people who live there.*

2. Work with small-scale farmers.
   • *Avoid the problems with massive, industrial farms and support local agricultural tradition.*
3. Use certified organic ingredients.
   • *Growing organic is a vital component of sustainability.*
4. Use fair trade ingredients (whenever possible).
   • *Giving farmers and workers a fair-share wage helps build communities and protect resources.*

Partnerships have been critical to Adina's social, environmental, and financial progress. By collaborating with various international organizations, Adina has been able to utilize the experience and resources of groups such as Agribusiness in Sustainable Natural African Plant Products (ASNAPP, http://www.asnapp.org), ECOCERT (http://www.ecocert.com), India's Eco Agri Research Foundation (http://www.ecoagri.in), and various co-ops involved with interests such as women's rights. Partnerships also contribute to a strong desire by Adina to open up the flow of information. Adina is a member of the Sustainable Food Laboratory and the Organic Trade Association (http://www.ota.com) (Figure 3.1). Steltenpohl believes, "Creating transparency in the supply chain is the single most important initial action any company can make towards sustainability."

Figure 3.1.   Adina's Senegalese hibiscus cooperative growers. (Photo courtesy of Adina for Life.)

Adina's global supply chain begins on the farm. In Senegal, Adina has built the supply chain up from scratch by first training and educating the hibiscus growers cooperatives about western business methods. At the same time this organizing work helped by aggregating volumes for more efficient sourcing and distribution. The initial groundwork was done by Adina who then handed the ongoing logistics over to ASNAPP who is still engaged and also acts as a marketing agent for the cooperatives. ECOCERT certifies the hibiscus as fair trade grown.

Adina does not try to specify what type of vehicle is used to move the product from farmer co-op to port since the logistics of this effort would be very complex and create a constraint and a burden of measurement on the cooperatives, and require resources that can better be used for sustainability initiatives elsewhere in the supply chain. ASNAPP currently handles the logistics of this distribution link, which adds more value and revenue to this Africa-based organization.

At the port, Adina contracts with DHL to provide freight forwarding services for the product containers, and DHL contracts with shipping companies to get the product on a boat and across the ocean to the United States. Adina then hires a truck to meet DHL at the port and bring the container to the processing center. Not all ingredients have required the same level of engagement. Sourcing fair trade coffee is relatively easy since it is such a ubiquitous commodity.

Adina is working to locate processing close to the farm sources of its global ingredients so that more jobs and money stay in these economies, thereby providing fair trade jobs in developing nations. Some of Adina's beverage ingredients are shipped as barrels of concentrated ingredients from their countries of origin. Steltenpohl calculates that this method not only ensures success on their fair trade goals but also reduces the carbon footprint of their finished goods. Shipping global concentrate to their California finishing plant can have more benefits than some locally sourced fruit juices. Financially, it costs less in nearly every circumstance. Environmentally, it allows ingredients to be produced where they grow best, and then be shipped via relatively more carbon-efficient sea transport; Socially, global sourcing significantly contributes to their initial goal of supporting traditional producers who grow traditional products.

Local distribution issues involve the manufacturing steps of their finished goods, so the carbon footprint of their Bay Area facilities is larger than for the agricultural and raw good transportation segments.

Steltenpohl says, "The biggest footprint is the last five miles." Adina has tried to locate its manufacturing and distribution centers close to their core Bay Area market. This is one reason why Adina purchased its own local distribution company in 2006.

Further, easily understood successes assist internal and external buy-in and show sustainability can be profitable. Adina has focused first on recycling practices: returnable pallets, packaging, and transportation materials that are significant material generators. Improved efficiencies, cost savings, and a decreasing carbon footprint have been the outcomes of this early work, complimented by purchasing carbon offsets for the materials for which they cannot realize a direct carbon reduction.

However, according to Steltenpohl, the biggest challenge that Adina faces is "to come up with a pricing structure that allows us to do the right thing in the supply chain at a price point that is acceptable to the market." This succinct statement is common to most businesses dealing with sustainability, for doing right by doing good can require costs that consumers may not be willing to pay since they do not see the value. Beyond this common constraint, Steltenpohl continually seeks transparency of processes and costs across all of its ingredients and practices.

National Distribution

This section of the chapter focuses on sustainable food distribution from a national perspective. The United States is the primary country surveyed for this national perspective; we will also use citations and examples from the United Kingdom.

Transportation is a major segment of the United States economy. According to the United States Department of Agriculture's Economic Research Service, roughly 4% of the marketing bill for U.S.-grown food is for truck and rail transportation (USDA-ERS, 2004). In 2005, transportation-related goods and services contributed $1.3 trillion to the $12.5 trillion United States gross domestic product. Americans spend a higher percentage per capita (18% vs 13%) on transportation than they do on food purchases (USDOT, 2007).

The United States possesses one of the most highly developed transportation infrastructures in the world. Table 3.2 indicates the extent of the United States transportation network in 2005.

Table 3.2.   The United States transportation network

Mode	Components
Road (highway)	46,873 miles of interstate highway 115,000 miles of other national highway 3,849,259 miles of other roads
Rail	95,664 miles of class I freight 15,388 miles regional freight 29,197 miles local freight
Air (jet or plane)	5,270 airports, including 26 large hub areas and 37 medium hubs
Water	26,000 miles of navigable channels More than 6,700 ocean waterway facilities More than 750 facilities on the Great Lakes 2,320 inland facilities

U.S. Department of Transportation (2007).

Transportation Modes and Impacts of Consolidation

In the United States, rail and water transport traditionally have been used to move lower-value agricultural commodities (such as corn, wheat, and soybeans). Rail is also used in some cases to move semi-perishable food items such as potatoes, squash, and onions. Air transport is used to move highly perishable and high-value food products. Transport by truck is the most common way to move food products, especially perishable items. Figure 3.2 shows the percentage of truck miles of various types of agricultural freight in the United States.

Over the past 25 years, the percent of food transported by truck in the United States has increased, while the percent moved by rail has decreased. This increase is highlighted in the arrival data collected from United States produce terminal markets. In 1981 approximately 50% of produce from all locations arrived at the Chicago terminal market by truck and 50% by rail. In 1998 nearly 87% of the produce arriving at the Chicago terminal market from all locations traveled by truck. Rail accounted for only 13% of the total produce arriving at the Chicago market in 1998 (Pirog et al., 2001). Truck transport is responsible for approximately 95% of fresh produce moving through the United States food supply chain; the remainder is by rail and piggyback (personal

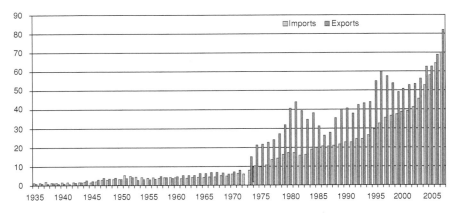

Figure 3.2. Percentage of total truck miles for food products carried in the United States.

communication with Camia Lane, USDA Agricultural Marketing Service Fruit and Vegetable Programs, January 24, 2008). In Great Britain, the amount of food moved by large trucks has increased 23% from 1978 to 2005, and the average distance for each trip has increased by more than 50% (Smith et al., 2005).

The United States infrastructure cannot easily accommodate the growing volume and shipper demands for food transport. Railroads carrying commercial freight are closing ramps and intermodal options, which puts increased pressure on the trucking industry to transport the United States food supply (Lemm, 2007). In 1965, there were 787,000 combination trucks registered in the United States; in 2004 there were more than 2 million (USDOT, 2007). United States carbon dioxide emissions from transportation have increased nearly 40% from 1985 to 2005 (USDOE, 2006). In addition to environmental concerns, the number of truckers remains in short supply. The American Trucking Association estimated in 2007 that the trucking industry will need to replace 539,000 truckers over the next 7 years. That means that approximately 54,000 new truckers will need to be hired each year to keep up with demand.

Consolidation of the retail food industry in recent decades has led to many changes. Large United States and European-based food retailers increasingly have purchased food products directly from manufacturers rather than wholesalers. Self-distribution has become the preferred method of supply chain coordination for large United States grocery stores. In 2001, nearly 82% of supermarket chains (those with 11 stores or more) used self-distribution to move products (American Institute of Food Distribution, 2003). This shift toward self-distribution among the larger grocery chains has led to increased bargaining clout among retailers over manufacturers. For smaller independent retailers that are not self-distributing, this means less bargaining power.

Consolidation at the purchasing end of the national food retail system is driving consolidation at the distribution level. Suppliers structure their operations to attain operating scales consistent with the needs of fewer, larger buyers. In the produce industry, for example, many grower/shippers have become multiregional and multicommodity in order to maintain a year-round presence in the marketplace.

The increase in self-distribution has led to a contraction in the number of United States food wholesalers. It is estimated that by 2012 the number of food wholesalers will decrease by more than two-thirds, eventually leaving only a dozen wholesalers to serve the national market (American Institute of Food Distribution, 2003).

Sustainability of the U.S. Food Distribution System

In a 2007 Food Marketing Institute-Harris poll of consumers, 92% of those polled agreed that it is important for the United States food industry (namely, food manufacturers and supermarkets) to be more proactive about addressing environmental concerns. Key environmental issues include fuel use and greenhouse gas emissions, air pollution, waste disposal and packaging, and total energy use.

An increasing numbers of European and United States retail and food service companies have integrated a strategy for sustainability into their businesses, including a reexamination of food transportation logistics. A report to the United Kingdom's Department of Environment, Food, and Rural Affairs (DEFRA), "Food Industry Sustainability," (2006) by Transport 2000, a U.K.-based consulting firm, indicated that a lower–carbon emission food system would have these features:

- Use of more seasonal and local produce local clustering (inputs to products would be sourced as locally as possible),
- Efficient management and operation of all processing
- Least use of temperature-controlled storage

- Logistical efficiency (such as increased fuel efficiency and consolidation of loads)

Specific food industry initiatives to improve vehicle utilization and efficiency include the following:

- Collaboration among operators
- Back-hauling of products, packaging, and waste
- New transport refrigeration technologies
- Using improved technology to optimize vehicle routing
- Improving vehicle full rates

An example of new transport refrigeration technology is *ecoFridge*, designed and manufactured in Europe. Unlike mechanical refrigeration systems, ecoFridge is nitrogen powered, eliminating greenhouse gas emissions and greatly reducing energy use.

Although greenhouse gas emissions are a key environmental challenge in national food distribution, there are other important issues that must be addressed to increase sustainability in food transport. The overall challenge is to provide efficient and profitable distribution of food and beverages to consumers while taking into account these associated environmental and social concerns:

- Air quality
- Climate change
- Traffic congestion
- Noise levels
- Waste reduction/reverse logistics
- Fair wages for workers involved in food transport

Research on the United Kingdom food industry in 2006 suggested that shorter supply chains that are logistically efficient can cut food transport emissions considerably (DEFRA, 2006). It is important to note, however, that a reduction in food miles (through sourcing of more local and regional food products) will not necessarily translate into reduced environmental impact. Mode of transport and transport vehicle fuel efficiency must be taken into consideration. A food mile is the distance food travels from where it is grown or raised to where it is ultimately purchased by the consumer or end user. A weighted average source distance (WASD) can be used to calculate a single distance figure, or food mile, that combines information on the distances from producers to consumers and amount of food product transported (Carlsson-Kanyama, 1997).

Pirog et al. (2001) examined United States Department of Agriculture's Agricultural Marketing Service produce arrival data from the Chicago, Illinois, terminal market for 1981, 1989, and 1998. A WASD was calculated for arrivals by truck within the continental United States for each year. Produce arriving by truck traveled an average distance of 1,518 miles to reach Chicago in 1998, a 22% increase over the 1,245 miles traveled in 1981 (Pirog et al., 2001).

Among the key environmental issues facing food distributors are the fossil fuel used and resulting greenhouse gas emissions released from food transport, as well as the resources required to maintain food quality and integrity during transport. Transportation accounts for approximately 14% of the energy flow (in British Thermal Units) in the United States food system (Heller and Keoleian, 2000). Consumers perceive that food traveling long distances will use more fuel and release more greenhouse gases than local food traveling shorter distances. In a 2007 national consumer study conducted by the Leopold Center at Iowa State University, two-thirds of respondents perceived that local produce grown in an open field in a neighboring county produced lower greenhouse gas emissions than produce grown elsewhere and then shipped cross-country (Pirog and Larson, 2007). On the contrary, a food product transported a longer distance by rail may actually use less fuel than a comparable product transported a shorter distance by truck because of the inherent fuel efficiency present with rail.

The following case studies highlight the sustainability efforts of two United States-based companies: SYSCO and Organic Valley. SYSCO is a distributor of products for restaurants and other food-serving institutions. Organic Valley is a producer and distributor of certified organic dairy, meat and produce products. Both these companies have national presence in the United States marketplace.

Case Study: SYSCO Corporation (http://www.sysco.com)
(Figure 3.3)

Background
SYSCO (Systems and Services Company) Corporation distributes a wide variety of approximately 400,000 food items as well as other food service products (850,000 stock-keeping units, or SKUs in all) to

Figure 3.3. SYSCO Corporation semitrailer. (Photo courtesy of SYSCO Corporation, Houston, Texas.)

restaurants, hotels, hospitals, and other food preparation entities. SYSCO has headquarters in Houston, Texas, and operates approximately 177 distribution companies with approximately 51,000 employees located across the United States and southern Canada. SYSCO's operations consist of four different types of distribution companies:

1. Broadline—providing more traditional food service options
2. SYGMA—specializing in chain restaurant distribution
3. Produce—focusing on produce distribution
4. Steak cutting—purveying fresh-cut steaks

SYSCO's base of 400,000 customers is composed of 64% independently owned restaurants, as well as chain restaurants, 10% health care institution; 5% schools and colleges; 6% hotels and motels; and 15% *other*, which includes theme parks, cruise ships, and camps. With fiscal year 2007 sales of approximately US$35 billion and delivering 1.2 billion cases of product, SYSCO represents 15% of the U.S. food service market.

SYSCO and Sustainability
In 2003, SYSCO's Chief Executive Officer appointed Craig G. Watson, who was then Vice President of Quality Assurance, to the additional role of SYSCO's first Vice President of Agricultural Sustainability. Several programs and initiatives to improve the company's sustainability have been developed in recent years, including the following:

- A *Buy Local, Sell Fresh* initiative was created in response to increased consumer demand for organic and natural foods.
- A Business Coalition for More Sustainable Food to research and improve social, financial, and environmental practices of their existing distribution networks focusing on key food business areas including transportation, distribution, and energy consumption.
- An integrated pest management program.
- An employee incentive program at the corporate and operating company management levels to reduce energy usage in the manufacturing and distribution of products.

SYSCO presently is conducting research on ways to improve the efficiency and sustainability of their distribution practices. Pilot projects include biodiesel, alternative fuels such as liquefied natural gas, fuel cells, and hybrid technology. SYSCO is currently working with ThermoKing to utilize compressed carbon dioxide as a refrigerant. The compressed carbon dioxide decreases emissions during transportation and reduces noise.

SYSCO is concerned about greenhouse gas emissions and climate change. The company closely examines and reevaluates all its delivery routes to optimize fuel efficiency and driver time and reduce greenhouse gas emissions. Early-morning delivery of loads enhances fuel efficiency, reduces traffic congestion and noise levels, and allows drivers to increase the number of stops made.

Rising energy costs impact distribution profitability and make it increasingly difficult for SYSCO to service more customers with smaller orders in less populated areas. It would be preferred if these customers could decrease the frequency of deliveries by increasing the size of each delivery, to reduce the amount of distribution time as well as increase its profit margin.

SYSCO is concerned with the increasing pressures of urban sprawl and water availability in high-volume, produce-growing regions in the United States, specifically in California. These environmental and land use concerns are forcing SYSCO to consider increased sourcing from Mexico and Chile. SYSCO is interested in determining whether existing growers in states with sufficient land and water availability can use technologies such as high tunnels to increase the length of the growing season and provide more product to one or more operating companies. Currently, since there is intense competition with retail outlets for produce items, SYSCO has moved to sourcing frozen broccoli and cauliflower from Mexico and changed its sourcing of blackberries from the Pacific Northwest to Chile. This has augmented availability of product, but has increased total food miles traveled.

SYSCO's customers, as well as the customers of most food distribution companies, want fresh produce 12 months of the year. Since much of this produce is not available in the United States year-round, this customer demand requires the company to source produce from countries such as Mexico and Chile. This situation increases distribution miles, and therefore increases fuel consumption and carbon emissions.

Case Study: Organic Valley (http://www.organicvalley.coop)
(Figure 3.4)

Background
Organic Valley Family of Farms was formed in 1988 in LaFarge, Wisconsin, with seven farmers. It has grown to a cooperative of more than 1,200 farm families—about 40% of United States organic dairy farmers and 10% of the nation's total organic farming output. Organic Valley has become one of the largest organic brands in the country, offering milk, cheese, juice, eggs, spreads, produce, and soy. Organic Valley's

Figure 3.4. Organic Valley Cooperative distribution center. (Photo courtesy of Organic Valley Cooperative.)

farmer–owners live in every region of the United States. This broad national spread of farms helps the company provide regional and local foods to its customers.

Organic Valley operations are based on two business models—the cooperative and the family farm. This allows the company to focus on providing quality food products using organic production practices. Farmer members of the cooperative have the opportunity to elect Organic Valley's Board of Directors as well as to participate in regional committees and directly influence the company's decisions. Organic Valley has established a profit-sharing model that allocates 45% of profits to farmers, 45% of profits to Organic Valley employees, and 10% of profits to the community.

Organic Valley and Sustainability

A primary goal at Organic Valley is to offset its ecological footprint through internal long-term measures of sustainability. The company hired a sustainability coordinator in 2007 and is developing a 3- to 5-year sustainability plan. Organic Valley plans to move beyond estimating greenhouse gas emissions to measure sustainability throughout its operations.

Organic Valley possesses the commitment and resources to implement its sustainability plan. The first phase of the plan, currently underway, will change the company distribution system so that it is more fuel efficient (using less total fuel and more biodiesel), allows more consolidation of products, and encourages more local and regional production, processing, and distribution.

Organic Valley's distribution model is set up to allow the production, processing, and sales of many of their products to be local and regional. For example, their dairies in the upper Midwest ship milk to 30 plants, keeping the delivery area as localized as possible.

Organic Valley also utilizes a third-party logistics company that combines truckloads with other organic and natural food companies, saving both fuel and time. This permits the company to use a cooperative transport model that makes efficient use of truck space by creating more full truckloads and back hauls. As part of its sustainability plan, the company is striving to achieve maximum efficiency through its transportation system. It is partnering with suppliers, customers, and vendors to use biodiesel produced under the most sustainable conditions available. Organic Valley is working with the Sustainable Biodiesel Alliance to develop principles that will serve as a basis for standards for a biodiesel certification process. These principles take into considerations greenhouse gas emissions, energy conservation, reductions of soil, water, and air pollution, and biodiversity conservation. Social principles such as food security, local consumption, and fair treatment of workers will also be considered.

Organic Valley is working with its farmer members to contract fuel oil crops grown using sustainable farming practices. In the future, more of the fuel used to power the distribution of organic products will come from its farmer members.

Organic Valley recently built a 100,000 ft^2 distribution center with an automated warehouse retrieval system. The warehouse is organized so they can vertically stack up to 12 pallets to increase energy efficiency. Organic Valley is also locating a biodiesel pumping station adjacent to this distribution center to provide contract carriers and farmer/members with the opportunity to use biodiesel. The 10 trucks owned by the company use 100% biodiesel fuel.

In order to promote sustainability in food production as well as environmental stewardship within their own organization, Organic Valley has created employee programs such as offering Organic Valley products at company cost, ride share and alternative commuting plans, and education initiatives that illustrate the importance of organic, local, and sustainable food products to employees.

Organic Valley continues to look for ways to partner with other organic companies so that it can ship full truckloads. The company continues to develop more efficient distribution system that allows contract haulers to use biodiesel as they come to pick up product. The ultimate goal is to create strategically located fuel sheds around distribution points so that all the product carried on truck uses biodiesel.

Regional/Local Food Distribution

Introduction

Local and regional food distribution depends on local and regional conditions and infrastructures. It is quite difficult to present a unified picture of sustainability issues in local and/or regional food distribution due to the profound variability of local conditions throughout the world. Even within the developed world, there are great variations in regional population distributions and distribution infrastructures, which present unique challenges. This section largely focuses on challenges faced throughout North America with regard to food distribution sustainability, although literature on the evolution of European markets will provide some European context.

Within North America and Europe, successive waves of advocates have voiced the need for more localized food systems, and their voices increasingly have been carried to the culture as a whole. New and revitalized direct relationships between many consumers and producers have resulted from this advocacy, with a strong resurgence in farmers' markets in North America and their widespread adoption in United Kingdom and Europe, along with the rise of community supported agriculture (CSA) enterprises. Farmers' markets in the United States have grown in number from 1,755 to 4,385 between the years 1994 and 2006 (USDA-AMS, 2007). There are now 1,306 CSA enterprises operating in the United States (Robyn Van En Center, 2008). Direct marketing sales as a whole (direct sales between producers and consumers like farmer's markets or CSAs) increased from US$592 million to US$812 million between 1997 and 2002 in the United States, from 110,000 to 117,000 farms (USDA-NASS, 2002). Given the continued growth of farmers' markets and CSA programs, there is every reason to expect even larger gains in the 2002–2007 period when the 2007 Census of Agriculture reports become available in early 2009. Direct marketing of specialty products has also been an important emerging opportunity for rural development in Europe over the past two decades in response to forces of globalization in major agricultural commodities (Marsden and Arce, 1995; Ilbery and Kneafsey, 1999; Marsden et al., 2000). Regional and process protection standards have also arisen throughout Europe as a way of supporting this specialty production and distribution, AOC, PDO, and PGI labeling schemes having received the most academic attention (Barham, 2002, 2003; Parrot et al., 2002; Ilbery et al., 2005). These standards regulate the labeling of specialty products according to regionally specific and distinctive production processes, the uniqueness of place through the notion of *terroir*, and unique qualities of the products themselves.

Direct marketing through farmers' markets, CSAs, and on-farm sales has not been the sole marketing manifestation of this movement. Local marketing also increasingly depends on direct farm sales to restaurants

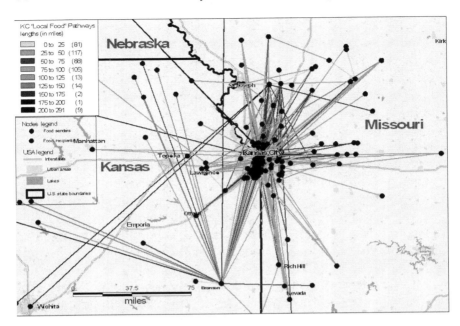

Figure 3.5. A map of regional/local food trade networks in eastern Kansas based on data gathered in the summer of 2005. The map shows food flows of a wide variety of fresh produce marketed and sold directly by producers to retailers and restaurants. Other food types in this region exhibited similar regional/local and urban/rural distribution patterns (Champion, 2007).

and retailers, circumventing traditional food processors and distributors. One of this chapter's authors mapped networks of local food sales in eastern Kansas for his doctoral dissertation during the summer of 2005 (Figure 3.5).

The food localization movement has linked itself implicitly with alternative forms of sustainable agriculture and a new wave of culinary culture centered on whole and heirloom/heritage foods. Yet, these agricultural and culinary alternatives are grounded in new supply chain geographies as a context for alternative producer–consumer relations. Notice the focus on urban marketing in the geographical distribution of these local food supply chains in Figure 3.5. This geographical pattern is the product of overlapping marketing relationships between hundreds of products and producers and their customers. Products vary from fresh produce to bottled milk, from processed bison snack sticks to native pecans, from fresh honey to fresh breads from local ingredients.

Participants include hobby farms, diversifying commercial farms, Anabaptist communities, small-scale food product manufacturers, small-scale abattoirs, high-end restaurants, supermarkets, specialty grocers, and more. This chapter's authors are most fluent in these developments in the United States, but literature indicates it is a growing trend in Europe as well (Marsden and Arce, 1995; Ilbery and Kneafsey, 1999; Marsden et al., 2000; Renting et al., 2003).

While the desire to remake food supply chains in terms of sustainability is a pervasive theme in the movement to localize food systems, there is great uncertainty about whether the local and regional food supply is contributing to more sustainable food systems. Small-scale agriculture and marketing may not be able to supply affordable food to current population levels without degrading soils and using too much land and resources. Accessibility and price of foods sold through local and regional supply chains may also present challenges to sustainability in terms of social justice. Merely in terms of liquid fuel consumption and carbon emissions, efficiencies of scale may prove that continental or even global supply chains use and emit less than uncoordinated localized markets. Even the primary criteria of sustainability may be a challenge for local and regional food supply, that of sustaining an adequate economic base for sustained operations and growth.

Local and regional food supply chains are often based on a reorganization of producer–consumer relationships; it is not so surprising that some of the most appropriate sustainability indicators for other scales have trouble capturing the challenges of the local/regional scale. This is especially true in the area of social sustainability indicators. The local and regional food supply chain sector in its current form is also rather young, full of diversity, and the source of much innovation. An attentive view to this emerging food system landscape can aid our abilities to formulate new sustainability indicators that better capture social and ecological relationships and their relevance of the notion of sustainability. We have chosen case studies at the local/regional scale, which highlight some of the more traditional sustainability indicators, but they also challenge our choice of indicators through their dependence on their unique respective local infrastructures and market demands.

In particular, a comparison of different CSA distribution models below leads to some conclusions about the importance of food miles as an indicator for sustainability when viewed through the lens of carbon emissions. Yet, it does not simultaneously compare the different economics and social impacts of these two distribution models, both from the producer and consumer–member perspectives. The other case study below of a retail-driven distribution model for a producer cooperative's products has its own limitations. This qualitative political economic analysis identifies strategic limits to growth and distribution challenges to sustaining alternative production and marketing. Yet, these limitations, and the efforts to transcend them, hide many of the costs to both people and the environment implicit in sustaining these local and regional business transactions. Emerging local/regional food supply chains face challenges of sustainability due to lack of standards, built infrastructure, and financing.

Case Study: CSA Food Miles

Recent research by Pirog and others at Iowa State University has compared the gasoline consumption and carbon emissions from two different CSA distribution models. The research is based on a case study of one farmer's CSA customers, and the comparison is between one model where the customers pick up their weekly allotment from a central location and another model where the farmer delivers the shares to member homes. The researchers calculate fuel consumption in the case of the central pickup model based on United States average fuel economy, while fuel consumption in the delivery model is calculated based on the average fuel economy for a Toyota Prius (the actual vehicle used by the farmer). The research results are partially represented in Figure 3.6.

The authors reported that most of the variation between fuel consumption in the two models is the result of different fuel economy for vehicles. The vehicle miles traveled between the two models are very similar, so fuel economy dictates fuel consumption in the comparison. The study shows that very significant fuel efficiencies can be gained—by both food distributor and consumer—simply by using more fuel-efficient transportation vehicles. The direct lesson from the study is that reducing total vehicle miles and maximizing vehicle efficiency are both very important as near-term strategies for minimizing the carbon footprint of local foods distribution. However, the results also suggest that personal transportation by consumers as a whole is a very prominent and often unconsidered force in carbon emissions in food supply chains. Finding ways that consumers could walk, bike, or at least carpool

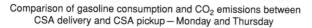

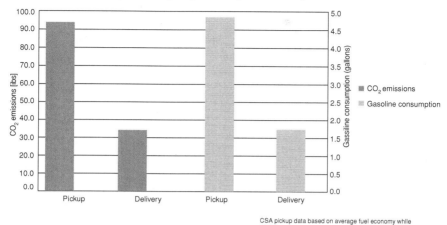

Figure 3.6. Comparison of gasoline consumption and CO_2 emissions between Iowa community-supported agriculture delivery and consumer pickup.

to pickup locations, shopping centers, or wherever they obtain their local foods will likely be an extremely important long-term sustainability strategy for local and regional foods distribution. Collaborations between farmers' markets and food retail stores to allow the market to be held in the parking lot of the food store could provide a reduction in total fuel use and increased overall sales of the farmers' market and the food store.

Case Study: Good Natured Family Farms (http://www. goodnatured.net/) (Figure 3.7)

Background

Good Natured Family Farms (GNFF) is a producer alliance among numerous independent farms as well as two meat producer cooperatives and a handful of small-scale artisanal food manufacturers. While the roughly 100 farms of the alliance are distributed throughout eastern Kansas and western Missouri (see Figure 3.8), it has sustained heavy growth in sales through a strategic relationship with a major grocery retail chain in the region's major urban center, Kansas City. The strength

Figure 3.7. Good Natured Family Farms products are featured at upper right in their retail-ready packaging. The scene is the typical retail environment for GNFF products, the Hen House suburban supermarket chain and its seasonal local food marketing campaign, complete with outdoor farmers market-like displays and robust internal local food signage.

of sales depends on a wide product variety, with products in every major grocery department, and a robust marketing brand and program that permeates the store atmospheres. Specifically, the program is a branch of the national *Buy Fresh, Buy Local* program managed by the Food Routes nonprofit organization, and the products must adhere to environmentally sensitive production standards as a condition of participation. Alliance members are inspected by an independent Kansas City environmental nonprofit organization according to these environmental production standards.

Distribution has consistently challenged the growth of this retail-driven marketing model, and current distribution focuses on incorporating GNFF products into a central Kansas City warehouse that serves only the grocery stores owned by GNFF's strategic retail partner. However, managing product supply to this warehouse from the dozens of product suppliers is quite a logistical challenge, and there are numerous product exceptions as well that must be delivered directly to stores.

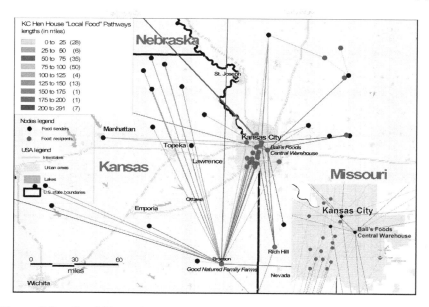

Figure 3.8. Good Natured Family Farms markets its products as *local* in Kansas City's suburban supermarkets, but it has actually constructed a regional supply network for a few dozen products spread among each of the major supermarket departments. Some of these supply chains directly connect farm to supermarket, while others require intermediary processing and/or distribution consolidation.

GNFF and Sustainability

GNFF has mainly focused on food production for its environmental sustainability efforts. Environmental oversight for the Buy Fresh, Buy Local program is administered by a local Kansas City environmental nonprofit through production standards that are verified with on-farm inspections. The standards are established in collaboration with the national Buy Fresh, Buy Local program managed by Food Routes.

GNFF has not dedicated much attention to environmental sustainability issues in their distribution model, but they have considerable experience in negotiating the structural limits of the region's food distribution infrastructure. The retail-driven strategic partnership allows GNFF to piggyback on the retail chain's investments in warehousing and distribution to share the economies of scale of these investments. This also provides uncommon access to professional distribution services for small-scale producers of artisanal products without requiring up-front investments or onerous distribution surcharges. Such a partnership has lowered the barriers to entry in the Kansas City retail market for the GNFF alliance, a key in building a wide membership and product diversity for the alliance, which in turn is critical for successful marketing in the stores.

GNFF has minimized its barriers to entry in medium-scale regional urban retail marketing, but it has paid a price in doing so. Its strategic relationship with an urban retail chain has allowed it to develop a robust marketing campaign in stores as the foundation for rapid sales growth, but there are limits to growth within this model. Ultimately, economic sustainability for the alliance requires continued sales growth, putting pressure on the strategic relationship that has been the foundation of their growth to this point. Furthermore, GNFF must confront the same barriers to entry in expanding its sales outside the strategic retail relationship as it did before the strategic relationship—the industry has not fundamentally changed.

First, the specialty food distribution industry is thus far ill-suited to handling small-scale specialty products. GNFF lacks a substantial marketing budget, and so any distributor they would work with must be able to act as a partner in generating sales through a combination of marketing and distribution. Specialty distributors are unaccustomed to performing this role for their customers, an approach that requires a shift from a commodity mentality to a quality and distinctiveness mentality. GNFF has experienced great resistance to adopting this mentality and form strategic partnership in its contacts with such specialty distributors. Such distributors also tend to shy away from sales and logistics at the scale of the case, rather preferring to manage product at the scale of the pallet.

Second, despite the opportunity to forge a new specialty distributor model, the scale of production and sales through GNFF has presented challenges to investing in its own distribution and marketing systems. Such investments would require organization and investments from alliance members that have proved difficult to coordinate. The alliance has explored establishing its own warehouse operation and transportation fleet to serve its products and members, an opportunity for socially equitable investments by all alliance members in environmentally sensitive energy systems for warehouse management and product transportation. However, there is a mismatch between the current scale of sales and the profit margins required for such infrastructure investments along with marketing expenses. It seems like there is no "just right" size for GNFF.

Case Study: Organically Grown Company
(http://www.organicgrown.com)

Background

Founded in 1978 by a group of gardeners, small-scale farmers, and environmental activists, Organically Grown Cooperative (OGC) has developed into the largest wholesaler of organic produce in the Pacific Northwest. Based in Eugene, Oregon, OGC has a staff of 135, operates 3 plants, 21 trucks, and 13 trailers, and markets the *Ladybug* brand of organic vegetables, fruits, and herbs.

In 1983 OGC formally aligned as a grower-owned cooperative, and recently transitioned to a grower-owned and directed S corporation. Today, OGC distributes to over 400 customers in the Pacific Northwest, including grocery store chains, natural food stores, co-ops, home delivery, restaurants, and buying clubs. In 2007 OGC generated revenue of US$60 million and distributed 2,706,477 cases. Fifteen million dollars was paid to the growers. OGC has grown over 20% every year since 1983 and today markets and distributes product from over 300 regional farms.

OGC and Sustainability

In 2005 OGC sponsored the newly formed Food Trade Sustainability Leadership Initiative organized by the University of Oregon's Resource Innovations. The goal of the initiative is to build the capacity of the organic produce trade to transition to sustainable business models, and to engage each leg of the supply chain in these efforts. The first Sustainability Summit determined that efforts should focus on three areas: (1) researching and sharing information about successful sustainable business strategies and practices in the food trade; (2) training practitioners to implement, measure, and report sustainable business practices; and (3) providing collaboration and networking opportunities across the organic and natural food trade supply chain.

In June 2005, Sustainability Coordinator, Natalie Reitman-White, teamed with the Oregon Natural Step Network to conduct seven, 3-hour-long sustainability orientations with the 109 employees who work in OGC's three Oregon and Washington facilities. The main task was to create a framework for sustainability and set measurable goals for the growth of sustainable efforts within that framework. The three goals that emerged from this work are discussed below, along with correlated performance metrics: (1) achieve carbon neutrality and eliminate fossil fuel use; (2) eliminate solid waste; and (3) support on-farm sustainability. From these goals emerged nine metrics for ongoing measurement: (1) waste output; (2) compost and recycling; (3) electricity usage and efficiency; (4) renewable energy percentage of total; (5) paper; (6) post-consumer fiber percentage of total; (7) fuel usage and efficiency; (8) renewable fuel percentage of total; and (9) packaging.

Some of the early work from these initiatives has involved efforts to highlight the best practices within transportation issues such as reducing idling truck time of company fleets to reduce emissions; installing fuel heaters to reduce the opportunity for biodiesel to resolidify (the majority of their fleet's biodiesel is sourced from spent fryer oil); and ongoing engagement across the supply chain to reduce the carbon footprint. Implementation of the emission reduction goal involved driver training about how to drive more efficiently, with an end result that drivers today are competing for lowest idle time and best mpg rating across their routes. OGC recently purchased one of the first hybrid electric heavy duty trucks that runs B20 biodiesel, a blend of 80% petroleum-based diesel, and 20% repurposed fryer oil (Figure 3.9).

Figure 3.9. Organically Grown Company's hybrid electric biodiesel truck (photo courtesy of Organically Grown Company).

Summary

Action on the first-tier issues with the largest known impacts generally results in early successes, which means more buy-in and revenue savings. Savings can then be made available to address second-tier issues. First-tier indicators, as highlighted in the Introduction, are more well defined, accepted, and measurable. These indicators appear across all scales of distribution—global, national, and regional/local—and include the following:

- Energy reduction, fuel efficiency
- Climate change, greenhouse gas emission reduction
- Profitability
- Efficient/reusable containers
- Transport efficiency (full load, backhauling)
- Engagement with other supply chain segments
  - Upstream/farmer efficiencies
  - Production standards (on-farm energy use, certification schemes)
  - Education
- Leadership, internal buy-in

Second-tier indicators include the following:

- Material/input substitution
  - Fuel conversion to biodiesel
  - Coolants used in refrigeration and freezing
  - Nonpesticide product stabilizers
- Employee incentive programs
- Consumer education

A third-tier redesign of global and national food distribution systems may include a thorough investigation of the economic, social, and environmental impacts of food products that are transported at these scales. Products that clearly have negative economic, environmental, and social impacts should be reevaluated for sustainability within local and regional scales. Likewise, local and regional products must be held to the same sustainability standards and reevaluated in national and global scales. Redesign may also take the forms of a shift away from eating certain food items out of season, a tax or tariff on total environmental burdens of a food product, or a significant shift toward food self-sufficiency and security for all nations.

The scale of an operation—global, national, or regional—does not appear to impede the ability of an organization to make measurable progress on first-tier indicators. Scale does, however, seem to influence the order of which sustainability initiatives are implemented, the uniformity of a starting point for initiatives, and the original motivation for initiating work toward an indicator. For some the motivation may be purely financial, while for others it may be an environmental or social issue tagged to the mission of the organization. Sustainable distribution initiatives have added benefits when done in concert with other components of the food industry. Integrating the supply chain into a sustainability initiative can maximize the impact, particularly on efficiency measures that work to reduce materials usage or emissions reduction, and aid the acceptance of new approaches to distribution. Supply chain education efforts, which may or may not involve end consumers, help build acceptance for innovative business approaches as well as understanding around the complexity of second-tier issues and why results on these indicators may take longer, are more expensive, or are not easily measured.

Organizations that have implemented sustainability initiatives have received internal and external benefits. Internally, measurable cost savings have been achieved by work on first-tier indicators, and the success generates excitement to pursue other efforts. Externally, engagement with other supply chain partners amplifies the gains that can be achieved by integrating sustainability initiatives across the supply chain and aid the development of best practices involving multiple indicators.

Ongoing work around sustainability in food distribution includes the refinement of best practices for many first-tier indicators and ongoing development of trade associations around specific issues that are best solved by the integration with supply chain partners.

References

American Institute of Food Distribution. 2003. Food Institute Report. Selected Issues. Elmwood Park, New Jersey.

Barham, E. 2002. Towards a theory of values-based labelling. *Agric. Hum. Values* 19:349–360.

Barham, E. 2003. Translating terroir: The global challenge of French AOC labelling. *J. Rural Stud.* 19:127–138.

Carlsson-Kanyama, A. 1997. Weighted average source points and distances for consumption origin-tools for environmental impact analysis. *Ecol. Econ.* 23:15–23.

Champion, B.L. 2007. *The Political Economy of Local Foods in Eastern Kansas: Opportunities and Justice in Emerging Agro-Food Networks and Markets.* Unpublished D.Phil. Dissertation, Oxford University, Oxford.

DEFRA. 2006. *Food Industry Sustainability Strategy 2006.* Report PB 11649. Available from http://www.defra.gov.uk/farm/policy/sustain/fiss/pdf/fiss2006.pdf. Accessed March 24, 2008.

Gerbens-Leenes, P.W., H.C. Moll, and A.J.M. Schoot Uiterkamp. 2003. Design and development of a measuring method for environmental sustainability in food production systems. *Ecol. Econ.* 46:231–248.

Heller, M.C., and G.A. Keoleian 2000. *Life Cycle-Based Sustainability Indicators for Assessment of the U.S. Food System (CSS00–04).* Center for Sustainable Systems. University of Michigan.

Hendrickson, M.K., and W.D. Heffernan. 2002. Opening spaces through relocalization: Locating potential resistance in the weaknesses of the global food system. *Sociol. Ruralis* 42(4):347–369.

Henson, S., and T. Reardon. 2005. Private agri-food standards: Implications for food policy and the agri-food system. *Food Policy* 30:241–253.

Ilbery, B., and D. Maye. 2005. Food supply chains and sustainability: Evidence from specialist food processors in the Scottish/English borders. *Land Use Policy* 22:331–344.

Ilbery, B., and M. Kneafsey. 1999. Niche markets and regional specialty food products in Europe: Towards a research agenda. *Environ. Plann. A* 31:2207–2222.

Ilbery, B., C. Morris, H. Buller, M. Kneafsey, and D. Maye. 2005. Product, process and place: An examination of food marketing and labelling schemes in Europe and North America. *Eur. Urban Reg. Stud.* 13(2):116–132.

Lemm, D. 2007. "Innovative truck/rail transport and future of trucking and agricultural shipping." In: *National Summit on Agricultural and Food Truck Transport for the Future, Washington, D.C. April 25–26, 2007.* Available from http://www.agandfoodtrucking.org/2007/downloads/Wed_DonnaLemm.pdf. Accessed March 15, 2008.

MacRae, R.J., S.B. Hill, J. Henning, and G.R. Mehuys. 1989. Agricultural science and sustainable agriculture: A review of existing scientific barriers to sustainable food production and potential solutions. *Biol. Agric. Hortic.* 6(3):173–219.

Marsden, T., and A. Arce. 1995. Constructing quality: Emerging food networks in the rural transition. *Environ. Plann. A* 27:1261–1279.

Marsden, T., J. Banks, and G. Bristow. 2000. Food supply chain approaches: Exploring their role in rural development. *Sociol. Ruralis* 40(4):424–438.

Parrot, N., N. Wilson, J. Murdoch. 2002. Spatializing quality: Regional protection and the alternative geography of food. *Eur. Urban Reg. Stud.* 9(3):241–261.

Pirog, R., and A. Benjamin. 2003. *Checking the Food Odometer: Comparing Food Miles for Local versus Conventional Produce Sales to Iowa Institutions.* Ames, Iowa: Leopold Center for Sustainable Agriculture.

Pirog, R., and A. Larson. 2007. *Consumer Perceptions of the Safety, Health, and Environmental Impact of Various Scales and Geographic Origin of Food Supply Chains.* Ames, IA: Leopold Center for Sustainable Agriculture. Available from http://www.leopold.iastate.edu/pubs/staff/consumer/consumer_0907.pdf. Accessed March 15, 2008.

Pirog, R., T. Van Pelt, K. Enshayan, and E. Cook. 2001. *Food, Fuel, and Freeways: An Iowa Perspective on How far Food Travels, Fuel Usage, and Greenhouse Gas Emissions.* Ames, IA: Leopold Center for Sustainable Agriculture.

Reardon, T., C.P. Timmer, C.B. Barrett, and J. Berdegue. 2003. The rise of supermarkets in Africa, Asia, and Latin America. *Am. J. Agric. Econ.* 85(5)1140–1146.

Reardon, T., C.P. Timmer, and J. Berdegue. 2005. *New Directions in Global Food Markets/AIB 794.* US Department of Agriculture-Economic Research Service. Available from http://www.ers.usda.gov/publications/aib794/aib794.pdf. Accessed March 15, 2008.

Regmi, A. 2001. *Changing Structure of Global Food Consumption and Trade.* Market and Trade Economics Division, Economic Research Service, U.S. Department of Agriculture, Agriculture and Trade Report. WRS-01-1. May 2001.

Regmi, A., and M. Gehlhar. 2005. *New Directions in Global Food Markets/AIB 794.* US Department of Agriculture-Economic Research Service. Available from http://www.ers.usda.gov/publications/aib794/aib794.pdf. Accessed March 15, 2008.

Renting, H., T. Marsden, and J. Banks. 2003. Understanding alternative food networks: Exploring the role of short food supply chains in rural development. *Environ. Plann. A* 35:393–411.

Robyn Van En Center. 2008. *CSA Database.* Wilson College, Chambersburg, PA. Available from http://www.wilson.edu/wilson/asp/content.asp?id=1567. Accessed March 15, 2008.

Schlich, E.H., and U. Fleissner. 2005. The ecology of scale: Assessment of regional energy turnover and comparison with global food. *Int. J. Life-Cycle Anal.* 10(3):219–223.

Smith, A., P. Watkiss, G. Tweddle, A. McKinnon, M. Browne, A. Hunt, C. Trevelen, C. Nash, and S. Cross. 2005. *The Validity of Food Miles as an Indicator of Sustainable Development.* DEFRA Report ED 50254. Available from http://statistics.defra.gov.uk/esg/reports/foodmiles/execsumm.pdf. Accessed March 14, 2008.

USDA-AMS (United States Department of Agriculture, Agricultural Marketing Service). 2007. *Farmers Market Growth.* Available from http://www.ams.usda.gov/farmersmarkets/FarmersMarketGrowth.htm. Accessed March 15, 2008.

USDA-ERS (United States Department of Agriculture, Economic Research Service). 2004. Amber Waves.

USDA-NASS (United States Department of Agriculture, National Agricultural Statistical Service). 2002. *Census of Agriculture*. Volume 1, Chapter 1, Table 2. Available from http://www.nass.usda.gov/census/census02/volume1/us/st99_1_002_002.pdf. Accessed March 16, 2008.

USDOE (United States Department of Energy), Energy Information Administration. 2006. *U.S. Carbon Dioxide from Energy Sources 2005 Flash Estimate*. Available from http://www.eia.doe.gov/. Accessed March 15, 2008.

USDOT (United States Department of Transportation), Bureau of Transportation Statistics. 2007. *Pocket Guide to Transportation 2007*.

Chapter 4

Packaging

Aaron L. Brody

Introduction

Packaging has served the role of keeping food safe and protecting it from loss and damage. Food waste in the United States represents about 15–20% of the initial agricultural production, the lowest in world history, and continuing to decline as a result of more efficien distribution infrastructure and its underlying packaging (Milgrom and Brody, 1975; Selke, 1990). Where food waste is high, a common occurrence in some developed and developing countries, hunger and even starvation are prevalent. The effective application of food packaging markedly reduces food waste. Thus, there are clear economic and social roles of packaging. Increasingly, packaging is being challenged to meet environmental goals, to provide more sustainable packaging options.

Environmental Considerations of Food Packaging

The life cycle and impacts associated with packaging have been evaluated. While the total impact of packaging tends to be very limited compared to other components of the supply chain, less than 2% for most food products, there are instances where the total impact of packaging on a food product's life cycle is significant up to 20%, due to the energy needed to produce the package or when packaging high water-containing products like beverages (Katajajuuri and Virtanen, 2007). The reasons for such an impact are that packaging uses nonrenewable resources, depletes natural resources, creates waste, and uses energy (Envirowise, 2008).

When looking solely at the package, it has been found that over 95% of the environmental impact is from the production of the package. The remaining 5% is in the disposal. A Tellus Institute study concluded that with the exception of polyvinyl chloride (PVC) plastic that has a significantly higher environmental impact, ". . . the lightest-weight package, per unit of delivered end product, is generally the lowest-impact product" (Ackerman, 1992). Further, the choice of container size may have more environmental burdens than either the choice of cup material or the cup manufacturing process. In a study conducted by the Center for Sustainable Systems at the University of Michigan, it was found that 32 oz yogurt containers consumed 27% less energy to produce and distribute than 8 oz containers (Brachfeld et al., 2001).

The drive to decrease material/resource use, waste, and energy use has been an aim, initially to decrease the cost of the package and the fina food product (Envirowise, 2008), but now have become the leading sustainability strategies for packaging.

Source Reduction to Minimize Resource Depletion

Resource use, especially of nonrenewable materials is one of the most significan issues for packaging's sustainability. It also influence the solid waste stream. As mentioned previously, the main driver for source reduction up until now has been cost reduction. Source reduction in the past has also been a result of aims for improved functional performance, for example, stand-up, fl xible pouches, and retort pouches.

Source reduction is typically done by light-weighting of the packages. This involves a conversion from a higher to a lower mass structure. Growing in practice is careful analysis of the package design, size, and shape to ensure the most efficien use of material. Biobased packaging is also emerging as an option to minimize resource depletion.

Light-Weighting

Package light-weighting during the past 30 years has led to the conversion of glass milk bottles to high-density polyethylene plastic that resulted in a mass reduction of 90%, and a 35% reduction of aluminum can weight for carbonated beverages and beer cans. Such material reductions are only successful, however, if the package continues to provide the protective characteristics needed for that food or beverage product. In some cases the distribution systems have been altered to accommodate to the package structure and performance change. During the 1970s introduction of polyester packaging like polyethylene terephthalate (PET) carbonated beverage bottles as replacements for glass bottles, channels were truncated to accommodate the reduced barrier properties of the plastic structures. Stand-up fl xible pouches began to be used in place of lined paperboard cartons or plastic jars when channels became better controlled relative to compression and relative humidity.

Transport costs for heavier packages are a growing limitation for some packaging options. Glass, which is widely recycled and made from recycled material, is being replaced more and more due to the cost of transporting the heavy material. Stoneyfiel Farm found that the energy (fossil fuels) used over the entire life of the glass package, for its manufacture and transport, exceeded the energy that goes into the manufacturing and transporting of a plastic container, and so they ended up using plastic packaging for their yogurt products (Stoneyfiel Farm).

Light-weighting can be achieved with new very low-mass package structures, such as multilayer packages. This is through lamination or coextrusion, or a combination. This mixed material, however, is impractical to recycle due to the high cost of separation of the material components. Some examples of multilayer packaging include the following:

- Metallized oriented polypropylene laminations for salty snacks
- Aseptic and extended shelf life paperboard/plastic laminations cartons for fruit beverages
- Barrier plastic ketchup, condiment, salad dressing, etc., bottles
- Aluminum foil/plastic lamination retort pouches for seafood, pet foods
- Aluminum foil/plastic lamination fl xible bricks for roasted and ground coffee
- Multilayer plastic fl xible pouches for cake mixes, cookies, and crackers

All these multilayer materials have represented major reductions in materials employed to protect food products in distribution channels. However, since they rely on nonrenewable materials and cannot be recycled, their benefit may be limited, especially when a life cycle analysis is conducted on the package.

This is why careful design of packaging to ensure effective use of materials has been growing. Recent examples of design for source reduction have proved to have several benefits For example, Unilever eliminated an outer carton from its Knorr vegetable soup mix and created a new shipping and display box that resulted in a 50% packaging reduction, 280 few pallets, and 6 fewer trucks a year to transport the same quantity of product (McTaggart, 2008).

Biobased Packaging

Light-weight and source-reduced packages have relied on nonrenewable materials. A potentially better design would be to produce such light-weight and source reduced packages from renewable resources (managed sustainably). As a result, significan work is underway to fin biobased packaging solutions.

Biobased materials by definitio conserve nonrenewable materials. They are materials extracted from biomass, polymers synthesized from biomass substances, or polymers produced biologically with microorganisms (Haugaard and Mortensen, 2003).

Biomass-derived packaging materials are typically cellulose or starch based. These end up being blended with hydrophobic polymers to reduce moisture migration. Polylactic acid (PLA) is an example of a polymer synthesized from biomass substrates. PLA is a lactic polymer, where the lactic acid is derived from organic products like corn. Poly(hydroxyalkanoates) (PHA) is an example of a polymer produced biologically with microorganism. Packages made from PHA are not widely available since they are expensive to produce (high cost of extraction and conversion) (Haugaard and Mortensen, 2003).

For most biobased packaging options, oxygen permeability is low, water vapor permeability is high, mechanical properties can be customized, and they may be biodegradable or compostable (Haugaard and Mortensen, 2003).

Biodegradation results in lower-molecular-weight fractions as a result of naturally occurring environmental microorganisms. These smaller pieces are further degraded by other means such as wind and rain. Photodegradation breaks polymers into small fragments as a result of exposure to visible and/or ultraviolet radiation. Composting involves biological degradation to carbon dioxide (a greenhouse gas) and water leaving no visible residual. The only relevant property for packaging is compostability, and only when the material is able to be diverted appropriately to be composted. Biobased materials are not automatically compostable, and some synthetic materials can be compostable.

PLA is the biobased packaging material that has received the most attention these days. PLA is synthesized from lactic acid monomers with lactic acid produced by fermentation of corn sugar or other sugars. The polymer ester linkages are sensitive to enzymatic and chemical hydrolysis as could occur in composting. PLA is commercial today, being extruded into fil   and sheets for food service containers and utensils and injection blow molded into bottles. PLA reportedly processes in a manner similar to polystyrene or polyester.

PLA has properties similar to polystyrene, but has been promoted to be near PET. PLA is thermally sensitive, softening at $<105°F$, thus confinin its applications to refrigerated cold fil products such as fresh-cut vegetables. Further, it has low moisture and gas barrier properties.

PLA cannot presently be composted in conventional municipal composting but rather requires special conditions. The major PLA resin supplier, Nature Works LLC, has a buy-back program for bulk waste PLA, which it then composts.

Economics of PLA today are based on a subsidized price of corn (which is rising because of its application for ethanol manufacture) and the elevated price of petroleum-based thermoplastics.

Currently, PLA-packaged food and food service applications include 100% PLA cups or PLA-coated paperboard cups; trays for organic foods; trays for relatively short shelf life refrigerated foods such as salads, egg cartons, yogurts, fresh-cut fruit, and vegetables. Bottled water has been packaged in PLA, but the high water vapor permeability of the material leads to bottle collapse during distribution.

A study reported in Scientifi American (Slater and Gerngross, 2000) indicated that PLA degradation produces greenhouse gases, CO_2, and CH_4, and that the quantity of energy to produce PLA is the same as that required for production of polystyrene for which it is substituting.

Packaging Waste and Waste Reduction

Food packaging is disposed of after the food product is consumed. While disposal of packaging is not a significan life cycle impact, given its visible waste, the issue of packaging's impact on solid waste is heightened. In the United States, almost 200 million tons of municipal solid waste (MSW) are generated annually. MSW is only a small fraction, about 5%, of total solid waste generated that includes sand and gravel, building debris, tires, coal slag, sewage sludge, and trees. Per capita solid waste

Sustainability in the Food Industry

Municipal solid waste by component (2004)

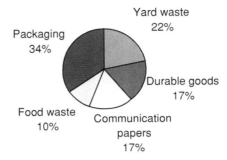

Figure 4.1.   Municipal solid waste by component. *Source*: Packaging/Brody, Inc.

in United States is about 4 lb, reflectin a slight decline in recent years. Figures 4.1 and 4.2 illustrate the composition of municipal solid waste in the United States.

In contrast to common belief, the overwhelming majority of *food packaging* solid waste is generated prior to home and food service operations. The most significan fraction of packaging waste is paperboard. This material includes corrugated fiberboar cases etc. can be recycled.

While the total waste that is composed of food packaging is small compared to all other sources of waste, it remains an opportunity to reduce the total life cycle impact of the food product. Reuse, recycling, composting, and use of edible packaging have been the primary means to reduce total waste from packaging. All these efforts have multiple

Waste by material

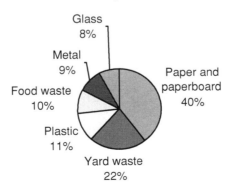

Figure 4.2.   Municipal solid waste by material (2004).

benefit in that they not only reduce solid waste, but may also provide a means for minimizing resource depletions.

Reuse
Reuse of packaging has been done effectively for limited types of packaging. Reuse has been done mostly for beverage and quick service food package materials largely because of their unit volume and visibility.

In reuse, the package is employed more than once. Reusable packaging today is widely used in Europe for beer. When, in the past, glass bottles were the primary packages, this system was applied in the United States for distribution of packages of beverages such as flui milk, carbonated beverages, and beer. To encourage return and reuse, deposits are typically used. Deposits function through the consumer paying some amount more for each bottle of beverage purchased, and that amount is returned when the bottle is returned. Such programs are still in act in the United States, in states such as Michigan and Oregon.

In general though, the delivery and pickup infrastructure used for primary packages such as glass bottles no longer exists in the United States. Reusable packaging is still used for food service draft (keg) and a small specialized niche of bottled beer multipacked in waxed corrugated fiberboar cases (which are not easily recycled at the end of their useful life). Corrugated fiberboar cases, injection molded plastic crates for carbonated beverages and flui milk, and plastic cases for fresh produce are used for distribution packaging to retail stores and returned for reuse.

Reusable plastic and corrugated fiberboar bulk packaging is increasingly used for internal food product transfer within and among industrial production facilities, for example, ingredients: syrups, powdered and granular goods, sauces, tomato concentrates, fruit concentrates, meat, poultry, as well as nonfoods.

Reusable packaging is necessarily underlain with issues of package sanitation: each container, even if used as a secondary package, may be a source of microbiological and possible pathogenic microbiological contamination and so must be cleaned and sanitized between uses.

Recycling
Recycling is the diversion of materials such as food package materials from the solid waste stream for application as raw materials in the manufacture of new products such as package structures.

Recycling today is dictated by availability of recovery of materials and market prices. The European Union Directive on Packaging and Packaging Waste from 1994, which harmonized national measures to manage packaging, included significan focus on recovery of packaging through recycling.

In closed-loop recycling, the material is returned for use on similar or identical package products. In open-loop recycling, the package materials are returned for use on any end product. Both end applications remove mass from the waste stream. However, energy is required to remove, return, and recycle the package. Further, recycling requires collection and sorting, which represents a cost. Separation is necessary since a few materials can be combined with other materials and then able to produce useful items.

Recycled materials come from different sources such as industrial, institutions, distribution channels, retail outlets, and consumers. Consumers recycle primary packages, after they have served their purpose. This recovered material is called post-consumer material. Post-consumer material has greater diversity and potential for contamination than any other source. As a result, post-consumer packaging ends up being the most expensive source of recovered material.

The main benefit of recovering/recycling food packaging is that it conserves resources, reduces emissions, has potential for energy conservation, may provide less expensive raw materials for products or packaging, and reduces solid waste. The main disadvantages are that the recovered material may be more expensive than the virgin material. Even though technologies exist for recycling almost all package materials, the availability of recovering the materials is the biggest challenge, which also results in unfavorable economics. Further, there is a reduction in functional properties of the material each time it is recycled. For example, recycled paperboard is inferior to virgin due to the shortening of the fibe (and thus strength) each time the material is processed, so a higher caliper (thicker) paperboard must be used to deliver the desired function. Additional considerations are the amount of energy needed to recycle the material (it may exceed the energy value from virgin materials), water usage, and potential for incorporation of contaminants that are possible toxicants.

Paper and Paperboard   Recycling of paper and paperboard has been a major technology for more than a century. As a result, it is the number one material recycled in the United States. Use of recovered paper and paperboard saves about 20–40% energy, compared to virgin material (EPA, 2008). However, as was mentioned, recycled paper has shorter fiber than virgin material. This reduces its functionality, and as previously mentioned, may require a thicker package to perform the desired function. Further, recycled paperboard usually cannot be applied as primary packaging because of the potential for contamination of contents and toxicity from the use of recycled material in primary packaging, that is, in contact with food contents. To illustrate use of recycled content, Kellogg uses over 90% recovered content in its cereal cartons (Envirowise, 2008).

Aluminum Recycling of aluminum has been a major and indispensable means of reducing costs of aluminum cans since the beginning of nonreturnable cans for beverage packaging during the 1960s. About two-thirds of aluminum cans are recycled back into aluminum sheet, which is then mostly used for the manufacturing of beverage cans. The success of aluminum recycling can be attributed to the economics of aluminum can recycling being favorable due to the high cost of electrical energy to refin aluminum from ore. Use of recovered aluminum saves over 90% energy, compared to virgin material (EPA, 2008). In aluminum recycling, the recovered aluminum material is shredded and inks and coatings are burned off during melting. Aluminum foil is a small fraction of the total, usually in laminations. Foil may be separated, but it is generally not economical to collect aluminum foil laminations.

Glass Recycling of postconsumer glass (after consumer use) is very involved because colored glasses must be separated, usually by manual means. The glass collected is composed of 65% clear (flint) 20% amber, and 15% green. There is little demand for green except for wine bottles in the United States. In spite of very high energy costs of melting glass, economics are unfavorable for postconsumer glass due to very high costs of returning scrap glass from consumer sites. Most of the reported recycling of glass is from inside glass bottle furnace and bottle/jar blowing operations.

Plastics Each food packaging plastic is different from each other, and so, for most potential end uses, each plastic must be separated. The major plastics mass is PET used for semirigid bottles and jars for carbonated beverages, salad dressing, water, etc. Additional plastic materials commonly used are high-density polyethylene largely used to make bottles for milk (and household chemicals and personal care items); polypropylene, mostly in fil   form; and polystyrene foamed and used for solid trays, for the many food service takeaway applications. Other plastics such as nylon and PVC represent very small fractions, and polycarbonate is not used for food or beverage packaging.

Only about 10–15% of PET is recycled in the United States, even in beverage bottle deposit states. Further, relatively few postconsumer plastics (of all types) end up being recycled.

At this point, there has not been enough recovered plastic to make it economically beneficia to not use low-cost virgin plastic material. But there is a significan energy savings when using recovered material (close to 80%) (EPA, 2008). The State of California has rigid plastic packaging regulations to encourage recycling and use of recovered content of such materials.

There have been food safety concerns with using recovered content in food packaging (and a United States Act and set of regulations to ensure safety). Developments have proved the feasibility of incorporating recovered content and as a result, it is practiced today. There are generally two means to incorporate recovered content into food packages, layering with virgin material so the virgin material is at the point of food contact in the package or high heat processing steps to *superclean* the recovered content (Dainelli, 2003).

Composting

The European Union has classifie composting as a form of recycling. Organic package materials such as paper and paperboard can be decomposed by aerobic microbiological decomposition. Other forms of biodegradable packaging are compostable, but are in negligible quantities today. An example is PLA. It does not compost well in municipal landfills Plastic food package materials contaminate composting. Note that composting is applicable mainly for true organics and for paper and paperboard package materials, which comprise about 40% of all food packaging. However, composting paper and paperboard is an inefficien application of this raw material, especially in the sustainability context. This is because degradation of organic materials generates greenhouse gases such as carbon dioxide, methane, and water vapor, and paper can be recycled for use in other paper products.

Edible Packaging

Edible packaging has an objective to potentially replace package materials from *nonrenewable* sources, that is hydrocarbon plastics. Sources used for edible packaging are plant-derived from annually renewable sources such as biopolymers, starch, cellulose, protein, lipids, and monomers from microbiological fermentation. Edible packaging waste is not necessarily compostable.

Edible packaging might function as adjunct to plastic package materials, enhancing quality and shelf life. It does not replace all conventional functional packaging. The advantages of edible packaging include the following:

- Can be consumed with the food product, leaving no visible residual for disposal (does not account for the sewage sludge that eventually is disposed of in landfills
- Degrades more rapidly than hydrocarbon polymers, paper and paperboard and, of course, glass, and metal
- Produced from renewable resources
- Might enhance product quality by incorporating fl vors etc.
- Can supplement nutritional value of food by incorporating nutrients etc.
- Can act as a barrier between components of heterogeneous foods to retard internal deterioration
- Can be carriers for active packaging such as antimicrobials and antioxidants
- Can incorporate microencapsulated food enhancers
- Direct contact layer in multilayer structures

Note that many of the benefit of edible food packaging are derived from adjuncts and additives that can also be incorporated into hydrocarbon polymers.

The key disadvantages of edible packaging are cost and limited functionality. Cost is especially an issue when the edible package must be functionally enhanced with other materials such as plasticizers and synthetic polymers to be rendered functional. In addition to the limited functionality of edible packaging, most edible packaging is hydroscopic.

Energy Use

Energy is needed to produce packaging and to distribute the packaging mass. This energy use can be minimized with source reduction strategies discussed. For example, reduction in package mass may reduce energy consumption in base material manufacture and distribution. Using recycled content may also provide energy savings (90% energy savings using recovered aluminum). Further considerations include reduction in energy consumption in downstream distribution (from refrigeration) as seen with aseptic and retort pouch packaging both significant y reducing and distribution refrigeration requirements (Robertson, 2006).

In addition, consideration should be given to the shapes of packages and the effect on transport efficien y, which effect total energy usage. For example, Envirowise suggests choosing packaging shapes that will maximize case and pallet utilization and transport efficien y, for example, rectangular sections and fla tops. In some cases it is worth considering stacking and nesting possibilities and shaping the pack accordingly to maximize palletization/transport efficien y.

Finally, use of renewable energy for packaging production could further reduce the total impact of energy used.

Design of Sustainable Packaging

Sustainable packaging has the goal of being beneficial safe, and healthy for individuals and communities throughout its life cycle. To meet this goal, packaging must meet market criteria for performance and cost; is sourced, manufactured, transported, and recycled using renewable energy; maximizes the use of renewable or recycled source materials; is manufactured using clean production technologies and best practices; is made from materials healthy in all probable end-of-life scenarios; is physically designed to optimize materials and energy; and is effectively recovered and utilized in biological and/or industrial cradle-to-cradle cycles (Sustainable Packaging Coalition, 2005).

Design of sustainable packaging can be achieved, or closely achieved, by considering the various approaches described to reduce the life cycle impacts of packaging by light-weighting, designing for reuse/recycling/

composting, and use of renewable resources, along with reducing waste and energy input. Further, food packaging already has made progress phasing out the use of toxics like those found in PVC (with phthalate plasticizers) and heavy metals (such as through U.S. regulations like *Plastic Materials and Articles in Contact with Food Regulations 1998*). Additional progress can be made to eliminate bleaching agents and solvent-based adhesives and fewer inks.

Sustainable packaging in these terms has recently been incorporated into an initiative from the major U.S. retailer Wal-Mart. Wal-Mart set a goal to reduce their overall packaging by 5%. To help enable this, in 2008, Wal-Mart launched a packaging scorecard system designed to evaluate packaging on the described sustainability considerations. The packaging scorecard is divided up accordingly (http://walmartstores.com/):

15%	Greenhouse gases/carbon dioxide tons of production
15%	Material value
15%	Product/package ratio
15%	Cube utilization
10%	Transportation
10%	Recycled content
10%	Recovery value
5%	Renewable energy
5%	Innovation

Conclusions

The role of packaging is to protect food in its distribution. This can enhance the sustainability of the supply chain (e.g., less waste). However, the life cycle impacts of packaging can be minimized by reducing the reliance on nonrenewable resources, minimize total resource use, minimizing total waste, and minimizing energy use.

References

Ackerman, F. 1992. *Tellus Institute Packaging Study*. Boston, MA: Tellus Institute.
Brachfeld, D., T. Dritz, S. Kodama, A. Phipps, E. Steiner, and G.A. Keoleian. 2001. *Life Cycle Assessment of the Stonyfield Farm Product Delivery System*. Center

for Sustainable Systems University of Michigan. Report No. CSS01-03. April 2001. Available form http://css.snre.umich.edu/css_doc/CSS01-03.pdf. Accessed September 15, 2008.

Dainelli, D. 2003. *Recycling of Packaging Maerials in Environmentally-Friendly Food Processing*. Cambridge, England: Woodhead Publishing Limited.

Envirowise. 2008. *Packaging Design for the Environment: Reducing Costs and Quantities*. UK: Envirowise.

EPA. 2008. *Puzzled About Recycling's Value: Look Beyond the Bin*. Available from http://www.epa.gov/epaoswer/non-hw/recycle/benefits.pdf Accessed September 15, 2008.

Haugaard, V.K., and G. Mortensen. 2003. *Biobased Food Packaging in Environmentally—Friendly Food Processing*. Cambridge, England: Woodhead Publishing Limited.

Katajajuuri, J.-M., and Y. Virtanen. 2007. "Environmental impacts of product packaging in Finnish food production chains." In: *Proceedings from the 5th International Conference on LCA in Foods, 25–26 April 2007*, Gothenburg, Sweden.

McTaggart, J. 2008. Seeing green. *Progressive Grocer* January 2008:44–67.

Milgrom, J., and A.L. Brody. 1975. *Packaging in Perspective*. Cambridge, MA: Arthur D. Little, Inc.

Robertson, G. 2006. *Food Packaging, Principles and Practices*. Boca Raton, FL: Taylor & Francis.

Selke, S. 1990. *Packaging and the Environment*. Lancaster, PA: Technomic.

Slater, S., and T. Gerngross. 2000. *How Green Are Green Plastics?* Scientifi American, Vol. 152, August.

Stoneyfiel Farm. *Stoneyfield Farm and Environmental Packaging*. Available from http://www.stonyfield.com/AboutUs/ ereDesigningForTheFuture.cfm. Accessed February 2008.

Sustainable Packaging Coalition. 2005. *Definition of Sustainable Packaging Version 1.0*. Available from http://www.sustainablepackaging.org/pdf/Definition 20of%20Sustainable%20Packaging%2010-15-05%20final.pdf Accessed September 15, 2008.

Chapter 5

Life Cycle Assessment across the Food Supply Chain

Lisbeth Mogensen, John E. Hermansen, Niels Halberg, Randi Dalgaard, J.C. Vis, and B. Gail Smith

Introduction

The environmental impact is one of the major pillars of concerns when addressing the sustainability of food production and sustainable food consumption strategies.

To assess to what extent food production affects the environment, one needs to choose a proper environmental assessment tool. Different types of assessment tools have been developed to establish environmental indicators, which can be used to determine the environmental impact of livestock production systems or agricultural products. The environmental assessment tools can be divided into the area based or product based (Halberg et al., 2005). Area-based indicators are, for example, *nitrate leached per hectare* from a pig farm, and product-based indicators are, for example, *global warming potential per kg pork* (Dalgaard, 2007). The area-based indicators are useful for evaluating farm emissions of nutrients such as nitrate that has an effect on the local environment. On the other hand, when considering the greenhouse gas emissions from the agricultural production, the product-based indicators are useful for evaluating the impact of food productions on the global environment (e.g., climate change) and have the advantage that in addition to emissions from the farms, emissions related to the production of inputs (e.g., soybean and artificial fertilizer) and outputs (e.g., slurry exported to other farms) are also included. In that way it is easier to avoid *pollution swapping*, which means that the solving of one pollution problem creates a new (Dalgaard, 2007).

Product-based evaluation is called life cycle assessment (LCA). LCA is an approach that evaluates all stages of a product's life. During this evaluation, environmental impacts from each stage is considered from raw material products, processing, distribution, use, and disposal. This methodology considers not only the flow of materials, but the outputs and environmental impacts of these. LCA processes have been standardized (e.g., ISO 14044) and follow the main steps of goal definition and scoping to define the process and boundaries; inventory analysis to identify material and energy flows and environmental releases; impact assessment to assess the environmental effects of the inventory analysis; and interpretation to draw conclusions from the assessment (SAIC, 2006). Conclusions can include decisions on different materials or processes. The benefit of LCA is that it helps avoid shifting environmental problems from one place to another when considering such decisions (SAIC, 2006).

Ultimately, the life cycle approach for a product is adopted to reduce its cumulative environmental impacts (European Commission, 2003). LCA is done in terms of a functional unit (FU)—for food that usually is a finished product like a pound of cheese or kg of meat. LCA has been used for environmental assessment of milk (Cederberg and Mattsson, 2000; Haas et al., 2000; Thomassen and de Boer, 2005; Weidema et al., 2007; Thomassen, 2008), pork (Cederberg and Flysjö, 2004; Eriksson et al., 2005; Basset-Mens et al., 2006; Dalgaard et al., 2007; Weidema et al., 2007), beef (Ogino et al., 2007; Weidema et al., 2007), grains (Weidema et al., 1996), and other agricultural/horticultural products (Halberg et al., 2006). Much of this information is included in the open access database LCAFood (www.LCAFood.dk), a comprehensive LCA database covering most food products produced under Danish/North European countries.

In LCA all relevant emissions and resources used through the life cycle of a product are aggregated and expressed per FU. Commonly applied environmental impact categories within LCA of food products are global warming, eutrophication, acidification, photochemical smog, and land use (Dalgaard, 2007). For each of the environmental impact categories, the emitted substances throughout the product chain that contribute to the environmental impact category are quantified (Table 5.1).

Global warming potential (GWP), the cause of climate change, refers to the addition of greenhouse gases to the atmosphere through burning

Table 5.1.   Selected impact categories with related units, contributing elements and characterization factors

Impact category	Unit	Contributing elements	Characterization factors
Acidification	kg SO_2 eq	SO_2	1
		NH_3	1.88
		NO_X^a	0.70
Global warming (GWP)[b]	kg CO_2 eq	CO_2	1
		CH_4	21
		N_2O	310
Eutrophication (nutrient enrichment)	kg NO_3^- eq	NO_X	1.35
		P_2O_5	14.09
		NH_3	3.64
		NO_3^-	1
		PO_4^{3-}	10.45
		NH_4^+	3.6
		COD^c	0.22
Land use	m^2	Land occupation	1

[a]NO and NO_2.
[b]Assuming a 100-year time horizon.
[c]Chemical oxygen demand: the amount of oxygen required to oxidize organic compounds in a water sample to carbon dioxide and water.
After Thomassen et al. (2008).

of fossil fuels, agricultural practices, and certain industrial practices leading to major changes in the earth's climate system. Nitrous oxide, methane, and CO_2 are the most important contributors to global warming, and, for instance, the contribution from agriculture to the Danish greenhouse gas emissions inventory has been estimated at 18% (Olesen, 2005). Nitrous oxide is emitted from slurry handling and from fields. For example, 4–5 kg nitrogen (N) from nitrous oxide (N_2O) per hectare per year is emitted from a typical Danish pig farm (Dalgaard et al., 2006), and although this is a small amount compared to ammonia and nitrate emissions, the contribution to global warming is significant, because nitrous oxide is a very strong greenhouse gas, 310 times stronger than CO_2. Methane is emitted from enteric fermentation, in particular from ruminant animals and from manure/slurry handling and storage. Fossil CO_2 is emitted from the combustion of fossil fuels (traction, transport, and heating). Finally, CO_2 can be emitted from the soil if more organic matter is degraded than build up in the soil.

Eutrophication is caused by the addition of excess nutrients to water. This results in algal blooms that lower the concentration of dissolved oxygen, and thereby killing fish and other organisms. Eutrophication contribution originates from a number of sources related to N and P emission on farm and handling of waste from processes after the farm. The N compounds include ammonia, which evaporate from the slurry in the stable, when the manure/slurry is stored, and after it is applied to the field. The ammonia can be deposited in vulnerable zones where it might decrease species richness because of eutrophication. Nitrate is another important N compound. Nitrate can be leached to the surface water or the groundwater; thus, it can cause both nutrient enrichment of the aquatic environment or pollution of drinking water.

Acidification is caused by release of acid gases, mostly from the burning of fossil fuels. Acid gas, for example, ammonia, has an acidifying effect and can affect natural habitats, some of which may be trans-boundary (e.g., lakes in Sweden). The major element that contributes to acidification from livestock production is NH_3 emitted from manure handling.

Production of food and animal feeds occupy some land that might have been used for other purposes eq maintaining biodiversity. The quality of the ecosystem is related to the biodiversity in the agricultural landscape. For example, soybean production for pig feed contributes approximately half of the total land use for pig meat. Increased soybean production results in agricultural expansion and causes a reduction in local biodiversity. However, land use is not only a negative concept, since part of the beef and milk production contributes to maintain valuable seminatural areas in the form of meadows (Weidema et al., 2005).

It is interesting to note that food production and consumption represent a large proportion of the total environmental impact that is related to human activities. In Table 5.2 the proportion of the impact categories is given (acidification, eutrophication, global warming, and nature occupation), which is related to the consumption of meat and dairy within the European Union (Weidema et al., 2007). While the total European consumption of meat and dairy products only constitutes 6.1% of the economic value of the total final consumption in Europe, meat and dairy products contribute from 14 to 35% to the impact categories like acidification, eutrophication, global warming, and nature occupation (Table 5.2).

Table 5.2.   Environmental impact of annual consumption of meat and dairy products in EU-27 (the functional unit of the study) expressed relative to the impact of EU-27 total consumption

Impact category	Unit	Impact of meat and dairy products relative to the total consumption
Acidification	m^2 UES	24.9%
Eutrophication, aquatic	kg NO_3 eq	29.4%
Global warming	kg CO_2 eq	14.2%
Nature occupation	m^2 arable land	35.8%

After Weidema et al. (2007).

These results highlight the importance of addressing the environmental impact related to food production.

Comparison of Environmental Impact of the Agricultural Production of Food Products

Food is thus an important component of the environmental impact of a family, but earlier assessments have demonstrated large differences in the environmental impact per kg product of different foods (Halberg et al., 2006). This is both because different products, such as milk and potatoes, obviously require different production processes, and because a particular product can be produced and processed in several different ways.

The potential environmental impacts associated with the on-farm production of various types of foods are shown in Tables 5.3 and 5.4. Producing 1 kg of animal products like meat and eggs produce much more greenhouse gas emissions than producing 1 kg plant-based product like potatoes. This is because the average amount of energy used per kg meat produced is more than 10 times that of plant-based products (Pimentel and Pimentel, 2003). For example, the energy from feed needed to produce 1 kg lamb meat requires 21 kg grain and 30 kg forage in feed input (Table 5.5), and the energy needed to produce 1 kg sheep meat is thus 23 MJ from animal feed, compared with 12 MJ for 1 kg of chicken meat and only 1.3 MJ for production of 1 kg of potatoes (Foster et al., 2006). Sheep and beef meat have the highest climate impact of all types of meat, with a GWP of 17 and 20.4 kg CO_2 eq/kg of meat, while pig and poultry have less than one-fifth of that (Table 5.3). Furthermore,

Table 5.3. The main burden and resources used arising from animal products from Denmark[a] or England/Wales[b]

Impact per kg carcass, per 20 eggs, or per kg milk at farm gate	Sheep[b]	Beef meat[a]	Pork meat[a]	Chicken meat[a]	Milk[a]	Eggs[b]
GWP$_{100}$ (kg CO_2 eq)	17	20.4	2.9	2.6	1.0	5.5
Acidification potential (g SO_2)	380	205	52	47	10.4	306
Nutrient enrichment (g NO_3 eq)	2,090	1,729	280	204	51	805
Photochemical smog (g ethane eq)	—	4.2	0.89	0.5	0.3	—
Land use (m^2 year)	14	31.5	8.9	4.9	1.5	6.7

[a]LCA Food (2008).
[b]Williams et al. (2006).

methane (CH_4) from enteric fermentation from cattle constitutes 32% of total greenhouse gas emissions from agriculture (Bellarby et al., 2008). So for ruminants, like sheep, beef, and dairy, methane production further increases greenhouse gas emissions per unit of food produced. Chicken meat production appears the most environmentally efficient due to several factors, including the very low overheads of poultry breeding stock (cf. 250 progeny per hen each year vs 1 calf per cow); very high feed

Table 5.4. The main burden and resources used arising from plant products grown in Denmark[a]

Impact per kg at farm gate	Bread wheat	Oilseed rape	Potatoes	Tomatoes (greenhouse)
GWP$_{100}$ (kg CO_2 eq)	0.7	1.5	0.16	3.5
Acidification potential (g SO_2 eq)	5.3	11.8	1.2	7.2
Nutrient enrichment (g NO_3 eq)	65	149	14	24.7
Photochemical smog (g ethane eq)	0.17	0.37	0.004	0.84
Land use (m^2 year)	1.5	3.5	0.31	0.02

[a]LCA Food (2008).

Table 5.5. Average consumption of grain/soy and forage (kg) for production of 1 kg of animal product (live weight)

	Beef meat	Pork meat	Chicken meat	Dairy (milk)
Grain/soy	3.5	2.6	2.0	0.4
Forage	38	0	0	1.8

LCA Food (2008).

conversion; and high daily gain of poultry (made possible by genetic selection and improved dietary understanding) (Williams et al., 2006).

The production of field crops produces much less greenhouse gas emissions than producing animal products (Table 5.4). The GWP from field crops (excluding protected cropping like tomatoes) is dominated by N_2O. N_2O contributes about 80% to GWP in wheat production (Williams et al., 2006). The N_2O contribution falls to about 50% for potatoes as much fossil energy goes into cold storage. In contrast, in greenhouse tomato production CO_2 from the use of natural gas and electricity for heating and lighting to extend the growing season is the dominant contribution to GWP (Table 5.4).

Comparison of Environmental Impact of Different Foods

For the consumers and the food industry, it is important to know the environmental impact of the produced food. The potential environmental impacts associated with various types of foods ex retail are shown in Tables 5.6–5.10. (All the foods are produced on farms in Denmark and processed in Denmark—http://www.LCAfood.dk)

Meat

The environmental impact of meat includes both the impact from the production of, for example, the living pig on the farm (Table 5.3), all the processes after the pig leaves the farmand until the meat arrives at the refrigerated counter in the supermarket. This includes the transport to the abattoir for slaughter, slaughtering, the cutting into primals, the packing, and the transport to the supermarket. However, the impacts associated with feed production, raising the livestock, and manure handling are the greatest contributor to the impacts noted.

Table 5.6. The potential environmental impacts of pork and cattle meat (functional unit is 1 kg food ex retail)

Impact per kg	Pork tenderloin	Pork ham	Pork minced meat	Cattle tenderloin	Cattle steak	Cattle minced meat
GWP_{100} (kg CO_2 eq)	4.6	3.0	2.3	68.0	42.4	4.4
Acidification potential (g SO_2 eq)	75	49	38	680	427	103
Nutrient enrichment (g NO_3 eq)	414	266	207	6,410	4,000	790
Photochemical smog (g ethane eq)	1.4	0.9	0.73	14	8.9	1.4
Land use (m^2 year)	12	8	6.0	90	56	11
2003 price in Danish supermarket (euro/kg)	18.8	12.1	9.4	40.2	25.5	9.4

http://www.lcafood.dk

Table 5.7. The potential environmental impacts of chicken and fish (functional unit is 1 kg food ex retail)

Impact per kg	Chicken fresh	Chicken frozen	Cod fresh	Cod fresh, fillet	Cod frozen, fillet	Shrimp fresh	Shrimp peeled, frozen	Mussels fresh
GWP_{100} (kg CO_2 eq)	3.2	3.7	1.2	2.8	3.2	3.0	10.5	0.09
Acidification potential (g SO_2 eq)	47.9	48.3	15	32	32	38	120	0.82
Nutrient enrichment (g NO_3 eq)	207	208	25	55	56	65	198	1.4
Photochemical smog (g ethane eq)	0.62	0.67	1.8	3.9	4.0	4.6	14	0.15
Land use (m^2 year)	5	5	—	—	—	—	—	—

http://www.lcafood.dk

Table 5.8. The potential environmental impacts of milk product (functional unit is 1 kg food ex retail)

Impact per kg	Skimmed milk	Low-fat milk	Full milk	Mini milk	Yellow cheese
GWP_{100} (kg CO_2 eq)	1.2	1.2	1.1	1.2	11.3
Acidification potential (g SO_2 eq)	12	12	11	12	101
Nutrient enrichment (g NO_3 eq)	58	56	53	58	467
Photochemical smog (g ethane eq)	0.42	0.42	0.40	0.43	3.3
Land use (m^2 year)	1.7	1.6	1.5	1.6	13
Fat content (%)	0.1	1.5	3.5	0.5	
Raw milk consumption (4.29% fat)	1.12	1.08	1.02	1.11	
Cream production (38% fat)	0.12	0.08	0.02	0.11	

http://www.lcafood.dk

Table 5.9. The potential environmental impacts of flour and bread (functional unit is 1 kg food ex retail)

Impact per kg	Wheat flour	Rye flour	Oat flakes	Rolls (fresh)	Rolls (frozen)	Wheat bread (fresh)	Wheat bread (frozen)	Rye bread (fresh)
GWP_{100} (kg CO_2 eq)	1.1	1.0	0.8	0.9	1.3	0.8	1.2	0.8
Acidification potential (g SO_2 eq)	6.9	6.8	7.0	5.1	5.4	5.0	5.3	4.9
Nutrient enrichment (g NO_3 eq)	84	73	17	59	60	59	60	54
Photochemical smog (g ethane eq)	0.34	0.39	0.40	0.29	0.32	0.27	0.30	0.29
Land use (m^2 year)	1.4	2.0	2.5	1.0	1.0	1.0	1.0	1.3

http://www.lcafood.dk

Table 5.10.   The potential environmental impacts of vegetables, sugar, and oil (functional unit is 1 kg food ex retail)

Impact per kg	Sugar	Vegetable oil	Potatoes	Carrots	Onions	Tomatoes
GWP_{100} (kg CO_2 eq)	0.96	3.6	0.22	0.12	0.38	3.5
Acidification potential (g SO_2 eq)	6.0	31	1.5	1.0	1.5	7.2
Nutrient enrichment (g NO_3 eq)	−12.1	439	14.4	3.6	15.0	24.7
Photochemical smog (g ethane eq)	0.83	2.1	0.14	0.15	0.15	0.84
Land use (m^2 year)	0.45	4.5	0.3	0.2	0.3	0.02

http://www.lcafood.dk

The environmental impacts associated with each class of pork/cattle meat have been determined by price allocation (Weidema, 2003) since it is anticipated that the most expensive cuts are major determinants of the *drive* in producing beef. This means, the total impact from producing one beef calf is divided among the output products, the different cuts according to the different prices. The resulting GWP for 1 kg cattle meat fluctuate from 68.0 kg CO_2 eq for tenderloin (the most expensive meat) to 4.4 kg CO_2 eq for minced meat (the cheapest meat). For pork, GWP fluctuate from 4.6 for tenderloin to 2.3 kg CO_2 eq for minced meat.

Meat is the food with the highest GWP (Table 5.6). The environmental impact from 1 kg meat from pig or poultry is of similar level (Table 5.7). For chicken the data shown are for uncut chicken and the processing including slaughtering etc. increased the GWP by 20%. However, if a frozen chicken instead of a fresh chicken is bought, the GWP is increased by additional 16%.

Fish

The fishing stage is the most important life cycle stage in terms of environmental burden for fish, and fishing activity is characterized by a significant fuel consumption and release of problematic biocides from antifouling paint on the boats (Thrane, 2003). The GWP and acidification potential of 1 kg wild cod fish is lower compared with 1 kg chicken, but dry matter content is not equal (Ellingsen and Aanondsen, 2006) (Table 5.7). The GWP of 1 kg fresh shrimp is similar to that of 1 kg fresh chicken, but three times higher for 1 kg peeled and frozen shrimps. Fresh mussels have a very low environmental impact for all impact categories. This is because mussels can filter plankton from the water and need no extra feed, and they can be raised on ropes hung on structures placed in coastal waters.

When looking at aquaculture, it resembles animal production more than fishing. As a result, the greatest impacts are typically seen in feed production (Ziegler, 2003). The relative environmental and resource use sustainability of aqua culture vs. fishing vs. livestock production needs further research.

Milk and Dairy

The environmental impact of different milk products is shown in Table 5.8. Manufacturing milk to drinking milk increases GWP by 19%. Further processing to shelf-stable milk (UHT treated) increases the processing impact due to higher heating requirements and energy to produce packaging (Hospido et al., 2007). The GWP of 1 kg cheese is 9 times higher than that of 1 kg drinking milk as it takes a lot of milk to make cheese (close to a 9×). It was assumed that milk is produced in a system without quotas and that milk fat in excess is converted into butter. Usage of milk from farm and contribution to butter production is specific for specific kinds of milk due to their different fat content (see Table 5.8).

Despite the need for refrigerated transport for most dairy products, agricultural production, including feed production, remains the key contributor to the life cycle environmental impacts (Hospido et al., 2007; Erzinger, 2003; Larsson, 2003). In particular, the feed has a significant contribution. Dairy production may be more of less integrated with land use and feed production and vary from zero grazing to totally free ranged systems. Thus, any attempt to estimate the effects of changes in the level of intensity in crop and livestock compartments of a dairy system or changes in feed composition should take into account, all relevant sub systems including the use of and the emissions and impacts from manure.

Organic dairy systems have been found to have lower emissions of GHG pr ha and per kg milk compared with conventional in Germany,

Sweden and Denmark, but not in the Netherland (Halberg et al., 2005, Thomassen et al., 2008). The nutrient losses were lower per ha and per kg milk in Denmark, Germany and the Netherlands.

Grain

The environmental impact of bread and flour is shown in Table 5.9. The processing cost of wheat into bread is an increase in GWP of 18%. However, if frozen bread is bought the GWP of the bread is increased by 43% compared with fresh bread. The bread is produced by baking dough made of flour and water and a number of other ingredients. The bread is baked in an industrial bakery using electricity for mechanical operations and light and natural gas for heating the oven etc. Compared with the animal products, the environmental impact of bread and flour is quite low; for example, the GWP of 1 kg bread is only 68% of that of 1 kg skimmed milk. It was demonstrated that organic wheat production could be more favorable than conventional production in relation to GWP (Braschkat et al., 2003; Nielsen et al., 2003).

Vegetables

Table 5.10 shows that field-grown vegetables and potatoes have a considerably lower GWP per kg product compared with other foods such as meat and bread. Both acidification and eutrophication for carrots, onions, and potatoes are less than 5% of the level of, for example, pork. Whereas tomatoes grown in a greenhouse have an environmental profile that is considerably different from field-grown vegetables, both because the heating needed in a greenhouse results in a relatively large emission of greenhouse gases and because the nutrient use is inefficient, leading to a significant loss of nitrogen and phosphorus (Halberg et al., 2006). A greenhouse production of vegetables is similar to a pork meat production in terms of greenhouse gas emissions, but causes slightly less acidification and nutrient loss. Compared with greenhouse vegetables, the field-grown vegetables have a low energy use and low emission of greenhouse gases per kg product, although it is somewhat higher for straw-covered carrots. The relative high importance of energy use in, for example, greenhouse tomatoes (heating) and carrots (soil preparation and straw coverage) and a large yield difference means that organic products have higher GHG emissions per kg compared with conventional.

The nutrient losses per kg tomatoes is, however, lower in organic production, which uses a soil based system rather than hydroponics.

Sugar and Oils

Table 5.10 also shows the environmental impact from sugar and oil. Sugar is produced from sugar beets produced in agriculture and transported by truck to the sugar factory where it is processed into sugar. The global warming and acidification potential from 1 kg sugar is similar to that of 1 kg flour, whereas photochemical smog is higher. Land use is lower due to a high crop yield in sugar beets, and when nutrient enrichment is negative, it is due to that molasses, and feed pills are cogenerated during sugar production and returned to agriculture as animal feed and thereby substitute grain feed.

Rapeseed oil is produced by crushing rapeseed. Rapeseed cake is coproduced with rapeseed oil in rapeseed crushing process, and these rapeseed cakes are used in animal feed.

Environmental Impacts of Different Meals

When comparing foods on a weight for weight basis, one should, however, be cautious as the products are not substitutable (they cannot completely replace each other) but rather complementary (we need a little of each) (Halberg et al., 2006). There is more protein, for example, in meat and dairy products than in vegetables and a different mix of vitamins. The environmental impact of the individual product must be seen in the light of how much it contributes to the total food consumption of a family. Table 5.11 shows a hypothetical evening meal for a Danish family, and Table 5.12 estimates the environmental impact of the meal. The meat has by far the largest impact. If you reduce the consumption of meat to approximately 100 g per person (the recommended level), you can reduce both the emission of greenhouse gases and the nutrient losses considerably (by respectively, 25 and 31%). The resulting environmental impact will, however, depend on whether you instead eat more field-grown vegetables or replace the meat with greenhouse vegetables. As Table 5.12 shows, half a kg of tomatoes will cancel out much of the saving in greenhouse gas emissions that results from a lower meat consumption (Halberg et al., 2006). It is also interesting to note that the

Table 5.11. Hypothetical meals for a family of four persons with different amounts of meat

	Typical	Less meat, more field-grown vegetables	Less meat, more greenhouse vegetables
Pork	0.75 kg	0.4 kg	0.4 kg
Potatoes	0.5 kg	0.75 kg	0.75 kg
Bread	0.5 kg	0.5 kg	0.5 kg
Milk	1.0 L	1.0 L	1.0 L
Carrots	0.5 kg	0.5 kg	0.4 kg
Onions	0.4 kg	0.2 kg	0.2 kg
Tomatoes	0 kg	0 kg	0.5 kg

Halberg et al. (2006).

environmental impact of the meal is considerably larger than the effect from, for example, driving a car for 3 km.

If the foods were ranked according to their environmental profile in the same way as in the classical food pyramid, some goods would switch places in the pyramid. Greenhouse-grown tomatoes, for example, would be at the top immediately below animal products. Field-grown vegetables would, however, be part of the staple diet, both from a nutritional and

Table 5.12. Environmental impact of different meals compared with that of transport, calculated using life cycle assessment method

Impact category	Typical	Less meat, more field-grown vegetables	Less meat, more greenhouse vegetables	Car (3 km)
Greenhouse effect (kg CO_2 eq)	4.4	3.2	4.0	1.1
Acidification (g SO_2 eq)	57	38	39	6
Eutrophication (g NO_3 eq)	315	218	223	8
Photochemical smog (g ethane eq)	1	1	1	5
Land use (m^2 year)	9.0	6.0	6.0	

Halberg et al. (2006).

environmental point of view, that is, at the bottom of the food pyramid (Halberg et al., 2006).

It is clear from the previous data that meat carries a huge environmental burden and that the on-farm part of the production chain is very important in this respect. Therefore, it is essential to consider the way the production takes place and to investigate if and where the environmental burdens can be alleviated.

Production Chain of Pork

When a pork chop reaches the refrigerated counter in the supermarket, it has accomplished a long journey as described by Dalgaard et al. (2007). First sows are raised to produce piglets; feed for the pigs is grown, harvested, and transported. Next, the pigs are fed, slurry is excreted, and then applied to the fields. The pigs are transported to the slaughterhouse, slaughtered, carved up, and finally the pork chop is brought to the supermarket, from where it ends up in the shopping basket of a consumer and finally on a dinner plate as illustrated in Figure 5.1. In each of these steps energy is used and pollutants are emitted. For example, artificial fertilizer is applied to the field where pig feed is grown and energy is used to produce this artificial fertilizer. In addition, different pollutants, for example, nitrate and nitrous oxide are emitted when the pig feed is grown or when slurry is excreted from the pig. Transport of fertilizer, pigs, and feed results in emission of CO_2 and other substances. All in all, many different kinds of pollutants in different amounts are emitted

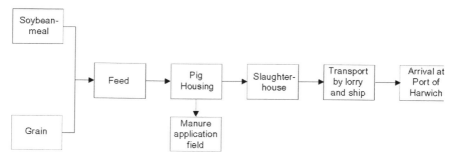

Figure 5.1.   Overview of the product chain of Danish pork delivered to the Port of Harwich in Great Britain. This represents a simplified view, where only the most important stages of the production chain are shown (Dalgaard et al., 2007).

Table 5.13. Results from different LCA studies on pork—1 kg carcass weight

Environmental impact	Global warming potential (kg CO_2 eq)	Eutrophication potential (g NO_3 eq)	Acidification potential (g SO_2 eq)
Farm gate			
Pork produced in Denmark (Dalgaard et al., 2007)	3.3	231	45.0
Pork produced in Great Britain (Dalgaard et al., 2007)	3.4	301	61
Pork produced organically in Denmark (Halberg et al., 2008)	3.8	353	75
Pork produced in Sweden (Cederberg and Flysjö, 2004)	2.6	170	37
Pork produced in France (Basset-Mens and van der Werf, 2005)	3.0	274	57
Pork produced in Great Britain (Williams et al., 2006)	5.6	760	290
From slaughter house			
Danish pork delivered at Harwich harbor (Dalgaard et al., 2007)	3.6	232	45.3

After Dalgaard et al. (2007).

before the pork chop is ready for consumption. These pollutants contribute to climate change, eutrophication (nutrient enrichment), increasing acidity in the aquatic environment, changes in biodiversity, or other undesired impacts on the environment.

Results from Different LCAs of Pork

A number of LCA inventories on pork have been performed—mainly focusing on the part until the farm gate. These results are summarized in Table 5.13. It appears that the results are quite consistent regarding GWP, whereas some variations exist among the other results from different sources. This may reflect differences in the situation analyzed, but, no doubt, methodological differences also exist.

In organic pig production, sows need access to grazing in the summer time and growing pigs need access to an outdoor run. A longer lactation period in organic system decreases the number of weaned piglets per sow per year, and a poorer possibility to adjust feed composition in the organic system results in a higher feed consumption per kg gain. When comparing Danish conventional (Dalgaard et al., 2007) and Danish organic pig production (Halberg et al., 2008), GWP is 12% higher for pig meat from the organic production system. However, the differences were larger for the eutrophication, where organic production is 52% higher, mainly due to leaching from the grasslands. Furthermore, the organic system had 67% higher acidification per kg pig meat due to larger ammonia losses from outdoor runs.

Global Warming Potential

The greenhouse gas emission per kg pork, carcass weight is 3.6 kg CO_2 eq. This equals the amount of greenhouse gas emitted from a 10 km drive in passenger car (LCA Food, 2008). The most dominating contributors to GWP are nitrous oxide, methane, and CO_2, and they are responsible for 44, 32, and 20%, respectively, of the greenhouse gas emissions (Dalgaard et al., 2007).

In Figure 5.2 the contributions to GWP from the different stages of the product chain of Danish pork are presented. The feed consumed by the pigs (*soybean meal* and *grain*) contributes with more than 2.4 kg

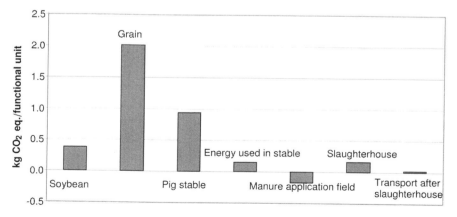

Figure 5.2. Contribution to GWP from the different stages of the product chain (Dalgaard et al., 2007).

CO_2 eq and is therefore more important than any other parts of the product chain (Dalgaard et al., 2007). The greenhouse gas emission per kg barley is 0.694 kg CO_2 eq (LCA Food, 2008), so with a feed use of 2.3 kg barley per kg pig live weight and 79.2 kg carcass weight per 105 kg live weight, the greenhouse gas emission from grain amounts to approximately 2 kg CO_2 eq/functional unit.

From *pig stable and storage* 81% of the GWP is methane and 19% is nitrous oxide. 78% of the emitted methane in the stable is from the manure/slurry and only 22% is from the enteric fermentation of the pigs. The nitrous oxide comes exclusively from the manure/slurry.

Contribution from *energy used in the stable* is both CO_2 emission from the production and distribution of electricity, and the CO_2 emitted from oil combusted for heat production at the farm. CO_2 is responsible for more than 98% of the greenhouse gases emitted. The contribution is 0.15 kg/functional unit and out of this 85% is from the use of electricity while the rest is related to the heat production from oil.

The contribution from *manure application field* is negative because less artificial fertilizer is used when the manure/slurry is applied to the fields for fertilization of the crops. The production and transport of artificial fertilizer emit greenhouse gases, so when artificial fertilizer is substituted by manure/slurry, greenhouse gas emissions will be reduced. On the other hand, when manure/slurry is used for fertilization instead of artificial fertilizer, more nitrous oxide will be emitted and more diesel for tractor driving will be used. However, the saved artificial fertilizer counterbalances more than this.

The contribution from *slaughterhouse* is 0.17 kg CO_2 eq/functional unit and is thereby the second smallest contributor to the GWP (Dalgaard et al., 2007). The major contributor from slaughterhouse is use of electricity at the slaughterhouse and the transport of the pigs from the farm to the slaughterhouse (distance 80 km). However, some of the by-products from the slaughterhouse cause saved emissions of greenhouse gases. Manure/slurry from the pigs is transported to a biogas plant where it is anaerobically digested, and the gas is used for heat and electricity production. The energy produced from manure/slurry substitutes fossil energy, and this results in a reduced emission of greenhouse gases. Also, the animal by-products (bone, blood, etc.) are used as bone and blood meal for animals or energy production. Nevertheless, the total avoided emissions of greenhouse gases due to manure/slurry and animal by-products are low, and in total they only sum up to −0.013 kg CO_2 eq/functional unit.

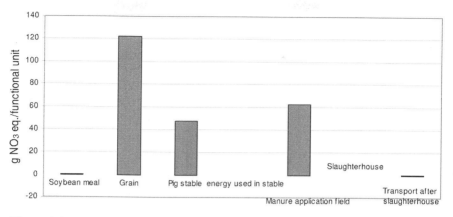

Figure 5.3. Contribution to eutrophication potential from the different stages of the product chain (Dalgaard et al., 2007).

Transport from slaughterhouse in Denmark to Harwich harbor is the stage of the product chain, which emits the smallest amount of greenhouse gases (Dalgaard et al., 2007). From the transport by lorry 0.021 kg CO_2 eq/functional unit is emitted and 0.007 kg CO_2 eq is emitted from the transport by ship. So even though the transport by lorry is only 126 km whereas the transport by ship is 619 km, the emission by lorry is three times higher. Less than 1% of the greenhouse gas emitted during the production of Danish pork can be ascribed to the transport from the slaughterhouse to Harwich harbor in Great Britain.

Eutrophication Potential

The most important contributor to eutrophication potential is nitrate (62%), followed by ammonia (32%), nitrogen oxides (4%), and phosphate (2%) (Dalgaard et al., 2008). As Figure 5.3 shows, the contribution from soybean meal is very low because nitrate, in general, is not leached during the cultivation of soybeans in Argentina (Dalgaard et al., 2007). The highest contribution to eutrophication potential comes from grain (122 g NO_3 eq/functional unit), with nitrate and ammonia emitted during the cultivation of the grain being the major contributors. The only contributing substance from *pig stable* is ammonia, which equals 47 g NO_3 eq/functional unit. The ammonia comes from the manure/slurry excreted in the stable and under storage. The contribution from *energy used in stable* is very low. The second highest contributor is manure application field, which contributes with 62 g NO_3 eq/functional unit.

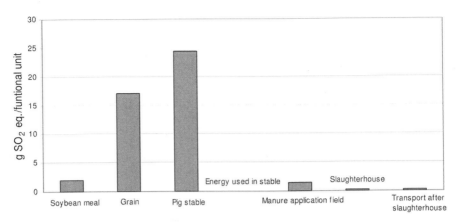

Figure 5.4. Contribution to acidification potential from the different stages of the product chain (Dalgaard et al., 2007).

A major part of this is N in the manure that is leached, because it is not incorporated in the crops. Slaughterhouse contributes with -0.4 g NO_3 eq/functional unit, and is negative because animal by-products, to some extent, are used as animal feed and thereby substitute grain feed. From the *transport after slaughterhouse* small amounts of nitrogen oxides are emitted due to fossil fuel combustion, but the contribution per functional is very low. The key element regarding eutrophication potential is N in the form of nitrate leached from fields and ammonia, emitted from the manure/slurry. The contribution from P is less than 2% per functional unit.

Acidification Potential

Ammonia is responsible for 84% of the acidification potential (Dalgaard et al., 2007). Nitrogen oxides, sulfur oxides, and sulfur dioxides, which come from the use of energy, are responsible for 16% of the acidification potential. The contribution from soybean meal is low (see Figure 5.4) and almost exclusively related to the emissions of nitrogen oxides, sulfur oxides, and sulfur dioxides emitted during the transport of soybean meal from Argentina to Denmark. Pig stable is the largest contributor to acidification potential, with ammonia as the only acidifying substance. Contributions from energy used in stable and slaughterhouse are very small. Manure application field contributes because ammonia is emitted when the manure/slurry is applied to the field. However, a significant part of that ammonia is counterbalanced because the manure/slurry applied to the field substitutes artificial fertilizer, which again results in saved emission from the use of fossil fuel.

Possibility for Environmental Improvement in Pork Production

The environmental *hot spots* in the product chain of Danish pork are, seen in relation to global warming, the stages before the pigs' arrival to the slaughterhouse (Dalgaard et al., 2007). A key parameter in reducing the GWP is farm management. If the protein consumption per pig produced is decreased, less N in manure/slurry will be excreted and thereby less nitrous oxide will be emitted from the pig stable. In addition, less protein consumption will result in a decreased use of soybean meal and a small increase in grain use. But because the greenhouse gas emission is lower per kg grain compared to soybean meal, a net decrease in greenhouse gas emission from the feed production will appear.

In the debate on climate change the focus is predominantly on CO_2 emissions and the use of energy by the industry and the transport sector. However, when considering food products (and in particular livestock products), methane and nitrous oxide are more important than CO_2 for the total impact on global warming. This is in accordance with the results presented above where the transport and the slaughterhouse are less important, but the emissions from feed production, stable, etc., are much more significant. But what if the Danish pork is transported to Munich in the south of Germany or Tokyo in Japan? To answer these questions two additional transport scenarios were established. One where the pork was transported 1,075 km by lorry (size 32 tons), which equals the distance from Horsens slaughterhouse to Munich and one scenario where the pork was transported 21,153 km, which equals the distance from Esbjerg harbor to Tokyo harbor in Japan. These longer transport distances increased the emissions from 3.6 kg CO_2 eq/functional unit to 3.7 and 3.8 kg CO_2 eq for the Munich and the Tokyo scenario, respectively. So even though the transport is much longer, the increase in the pork's contribution to GWP is limited.

How Unilever Uses Sustainability Tools and Metrics

Many definitions of sustainability can be found. In Unilever we believe that sustainability equates to the triple bottom line principle: any activity that adds to economic, social, and natural capital is more sustainable. This principle goes beyond limiting negative impacts: For any

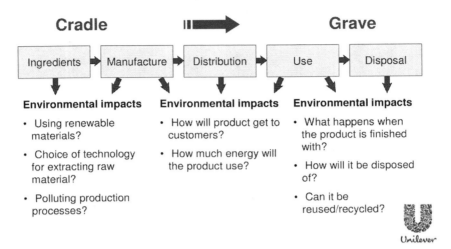

Figure 5.5. Scope of LCA.

organization to be able to grow in a sustainable way, its impacts on economic growth and on social progress and on natural resources must be positive.

We believe that we need to monitor and measure our sustainability impacts, both to help ourselves understand our own business and where interventions are most appropriate and to help us communicate our progress. We have therefore been involved in life cycle assessment and in developing and piloting other sustainability metrics for many years.

Life Cycle Analysis

Many aspects of the environmental sustainability of a food supply chain, particularly those linked to processing, manufacturing, transport, use, and disposal, can be assessed using life cycle analysis (LCA). LCA is a tool for evaluating the environmental burdens associated with a product, process, or activity over its entire life cycle (Figure 5.5).

In order to do this, a functional unit needs to be defined, and system boundaries described. This is called the scoping phase. For this production system that will produce the functional unit, for example 1 ton of canned tomato soup, all separate process steps are described. For instance, when looking at the production of a certain volume of a product, all inputs and outputs of the production process will be included. Also, energy and water use will be assessed. Packaging material will be taken into account, and all this from *cradle to grave*, that is, from the growing

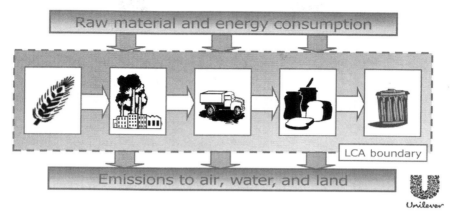

Figure 5.6. LCA boundary.

or mining of the raw materials until the waste phase after (consumer) use. For each process step, all inputs and outputs are collected in a quantitative way (see Figures 5.6–5.8).

The objective of the assessment is to translate the quantitative input/ output information that came out of the inventory into quantitative information on environmental impacts. Some compounds emitted to air contribute to global warming (climate change). Some compounds

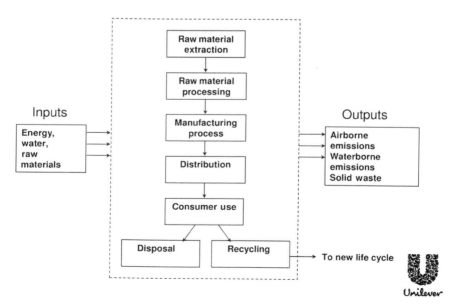

Figure 5.7. Process diagram.

Energy and raw materials		Solid waste	
Coal	1,500 kJ	Mineral	175 g
Oil	1,000 kJ	Slag/Ash	8 g
Gas	400 kJ	Plastics	17 g
Bauxite	20 g	Glass	2 g
Sand	120 g	Metals	25 g

Airborne emissions		Water-borne emissions	
CO	0.5 g	COD	4 g
CO_2	585 g	BOD	1 g
NO_x	3 g	Susp. solids	12 g
SO_x	6 g	Cl^-	0.2 g
Cl_2	0.1g	Organics	1 g
VOC	97 g	Metals	0.1 g

Unilever

Figure 5.8.   Life cycle inventory.

contribute to ozone depletion, or acidification, or both, or all three. Impact assessment therefore consists of a number of consecutive steps: classification, characterization, and categorization (see Figure 5.9). The classification step determines which compound contributes to which environmental impact category (e.g., climate change and ozone depletion). The characterization step quantifies this impact in certain predetermined units (e.g., GWP is expressed in CO_2 equivalents). The categorization step then adds the individual contributions of the individual emissions together to a total impact potential (Figure 5.9). Impact categories often included in LCA are listed in Table 5.14.

The result of the impact assessment is a so-called eco-profile (see Figure 5.10). The eco-profile gives a quantitative view of the potential contributions; the system or systems under study will have on the environmental impact categories that were defined in the scoping phase of the study. Sometimes the eco-profile will give absolute numbers, sometimes, as in the example in Figure 5.10 when comparing different production systems, alternatives will be normalized against each other.

If in the case of comparing two alternatives, the resulting eco-profile shows that for each individual environmental impact category contribution of alternative A is higher than that of alternative B—the interpretation is simple. From an environmental impact point of view, alternative

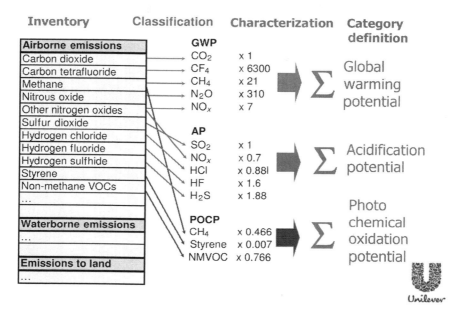

Figure 5.9. Impact assessment.

B is better. So all other things being equal (e.g., no difference in costs and investments, no difference in social impacts such as employment), in a green field situation, the choice will have to be for option B. But this is rarely the case. Almost invariably, one alternative will score better on some impact categories, and the other alternative will score better on the rest. And then the question becomes: how much improvement in GWP equals, how much improvement in ozone depletion potential? In other words, the LCA practitioner is always comparing apples and oranges.

The eco-profile might be used to identify options for improvement. By focusing on where contributions to environmental impact categories are largest, and examining where in the production system these contributions originate, it can become clear what improvement options (process

Table 5.14. Impact categories in LCA

Energy consumption	Resource depletion
Global warming	Ozone depletion
Eutrophication	Water consumption
Photochemical smog	Land use
Acidification	

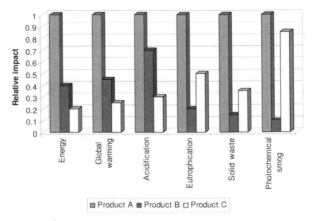

Relative comparison

Figure 5.10.   Eco-profile.

improvement, changing type of input or raw material, changing type of process, etc.) are available. Since the methodology is data intensive, and since there are no exhaustive databases (yet?) with this type of process data, the LCA process usually forces the researcher to get in touch with other players in the production system and ask questions that otherwise would not have been asked. It helps to do this type of project with a consortium of players, simply to allow a better understanding to emerge amongst all actors in a production system of what impacts downstream can be caused by certain decisions or conditions that are taken or exist upstream.

In all cases, executing an LCA is a great learning experience. The methodology forces the practitioner to think through what the production system is, how the elements in the production system are linked together, what the different outputs are (main product and by-products), what environmental impacts are relevant for the production system in question, and what options for improvement there might be. It will not, however, lead to definitive answers. If only because there are so many assumptions that need to be made (on where the system boundary is, on how impacts are allocated over various products) and because the data will always be incomplete. Where standard databases do not provide an answer, one has to look for alternative sources or make an approximation. And where the standard databases do provide an answer, the question is always whether your suppliers (your partners somewhere in the production system) are actually better or worse than the (average) numbers in the standard database.

A typical LCA result for a food product will often show the biggest impacts upstream in agriculture. The high impacts often result from the use of farm inputs (fertilizer, crop protection agents), which usually have a fairly high energy content and partially end up in the environment (either soil or ground and surface water). Transport is not an important part of energy consumption over the product life cycle, as a rule. Shopping by car can have a higher impact than long-distance freight transport, either by road or by sea, although mode of transportation has an effect of course (transport by ship and rail being more environmentally benign than transport by road).

A typical LCA result for a home care (detergent or hard surface cleaner) or personal care product (shampoo, shower gel, toothpaste) will show the biggest impacts downstream in the consumer use phase. This results from the fact that consumers usually have to heat water to use the product (for washing clothes, for taking a shower) and that the product, while used, immediately turns into aqueous waste.

This is why Unilever, with both food and home and personal care products, has both a sustainable agriculture and a sustainable water program.

Conclusions

From the food product life cycle research conducted globally, agriculture production is generally the largest contributor to the life cycle impact compared with other compartments such as transport and processing. Further, animal products have greater impact than plant products—producing 1 kg of animal products like meat produce much more greenhouse gas emissions than producing 1 kg of plant-based products like cereal or potatoes. This is due to the animal feed conversion rate and feed impacts themselves and to the emissions of nutrients and GHG from the livestock.

However, certain ways of production can increase plant product's impact, as was demonstrated with greenhouse growth of tomatoes being similar in impact to animal products. Organic production is most often more energy efficient and has lower GHG emissions compared with conventional while nutrient losses are lower per ha but not always per FU. The supposed environmental benefits of non-use of pesticides in organic systems are usually not included in LCA's due to methodological difficulties. Thus, comparing the two systems using State-of-art LCA is not fully satisfactory. Besides this, there is large variation in environmental impact between farmers and farming systems producing the same livestock output. LCA methodology may be used to benchmark the better performing systems and product chains in and to demonstrate the relative importance of the feed production external to the livestock farm itself.

Downstream compartments have relatively lower impacts, but can range depending on the product. Even more important, the relative high proportion of food wasted in households adds significantly to the environmental burden per kg of food actually consumed. Consumer transport to purchase food can be a significant impact. And finally, consumer use of the food, when including cooking, can be a major contributor to the life cycle impact. In general, following to a high degree current health advise regarding diet composition, especially eating a high proportion of basic vegetables will also minimise the environmental impact per meal. Thus, changing diets are potentially one of the most powerful ways of reducing the environmental impact per capita.

References

Basset-Mens, C., and H.M.G. van der Werf. 2005. Scenario-based environmental assessment of farming systems: the case of pig production in France. *Agric. Ecosyst. Environ.* 105(1–2):127–144.

Basset-Mens, C., H.M.G. van der Werf, P. Durand, and P. Leterme. 2006. Implications of uncertainty and variability in the life cycle assessment of pig production systems. *Int. J. Life Cycle Assess.* 11(5):298–304.

Bellarby, J., B. Foereid, A. Hastings, and P. Smith. 2008. *Campaigning for Sustainable Agriculture*. Amsterdam, the Netherlands: Green Peace International.

Braschkat, J., A. Patyk, A. Quirin, and G.A. Reinhardt. 2003. "Life cycle assessment of bread production—a comparison of eight different scenarios." In: *Life Cycle Assessment in the Agri-Food Sector: Proceedings from the 4th International Conference, October 6–8*, Bygholm, Denmark.

Cederberg, C. 2003. "Life cycle assessment of animal products." In: *Environmentally-Friendly Food Processing*, eds B. Mattsson and U. Sonesson. Cambridge, England: Woodhead Publishing Limited.

Cederberg, C., and A. Flysjö. 2004. Environmental assessment of future pig farming systems—quantifications of three scenarios from the FOOD 21 synthesis work. *SIK-rapport* 723:1–54.

Cederberg, C., and B. Mattsson. 2000. Life cycle assessment of milk production—a comparison of conventional and organic farming. *J. Clean. Prod.* 8:49–60.

Dalgaard, R. 2007. *The Environmental Impact of Pork Production from a Life Cycle Perspective*, p. 135. Ph.D. thesis, Aalborg University, Denmark.

Dalgaard, R., H. Halberg, I.S. Kristensen, and I. Larsen. 2006. Modelling representative and coherent Danish farm types based on farm accountancy data for use in environmental assessments. *Agric. Ecosyst. Environ.* 117(4):223–237.

Dalgaard, R., N. Halberg, and J.E. Hermansen. 2007. Danish pork production. An environmental assessment. *DJF Anim. Sci.* 82:1–34.

Dalgaard, R., J. Schmidt, H. Halberg, P. Christensen, M. Thrane, and W.A. Pengue. 2008. LCA of soybean meal. *Int. J. LCA.* 13(3):240–254.

Ellingsen, H., S.A. Aanondsen. 2006. Environmental impacts of wild caught cod and farmed salmon – a comparison with chicken. *Int. J. LCA.* 1: 60–65.

Eriksson, I.S., H. Elmquist, S. Stern, and T. Nybrant. 2005. Environmental systems analysis of pig production—the impact of feed choice. *Int. J. LCA* 10(2):143–154.

Erzinger, S., D. Dux, A. Zimmermann, R.B. Fawaz. 2003. "LCA of animal products from different housing system in Switzerland: Relevant of feed stuffs, infrastructure, and energy use." In: *Life Cycle Assessment in the Agri-Food Sector: Proceedings from the 4th International Conference, October 6–8, 2003*, Bygholm, Denmark.

European Commision. 2003. *Communication from the commission to the council and the European parliament. Integrated product policy.* Building on Environmental Life-Cycle Thinking. COM(2000) 302 final. Available from http://www.aeanet.org/forums/IPPfinalcommunication18June2003.pdf. Accessed June 2, 2008.

Foster, C., K. Green, M. Bleda, P. Dewick, B. Evans, A. Flynn, and J. Mylan. 2006. *Environmental Impacts of Food Production and Consumption: A Report to the Department for Environment Food and Rural Affairs*, pp. 1–199. Defra, London: Manchester Business School.

Haas, G., F. Wetterich, and U. Geier. 2000. Life cycle assessment framework in agriculture on the farm level. *International Journal of LCA* 5:345–348. Available from http://www2.dmu.dk/1_viden/2_Publikationer/3_fagrapporter/rapporter/FR380_samlet.pdf (in Danish).

Halberg, N., R. Dalgaard, and M.D. Rasmussen. 2006. Miljøvurdering af konventionel og økologisk avl af grøntsager: Livscyklusvurdering af produktion i væksthuse og på friland: Tomater, agurker, løg, gulerødder. Arbejdsrapport fra Miljøstyrelsen 5.

Halberg, N., J.E. Hermansen, I.S. Kristensen, J. Eriksen, and N. Tvedegaard. 2008. "Comparative environmental assessment of three systems for organic pig production in Denmark." In: *Proceedings of ISOFAR Conference: Organic Agriculture in Asia*, Korea, March 13–14, pp. 249–261.

Halberg, N., H.M.G. van der Werf, C. Basset-Mens, R. Dalgaard, and I.J.M. de Boer. 2005. Environmental assessment tools for the evaluation and improvement of European livestock production systems. *Livest. Prod. Sci.* 96:33–50.

Hospido, A., M.T. Moreira, G. Feijoo. 2007. "Comparative LCA of liquid milk: Pasteurized versus sterilized milk." In: *Proceedings of the 5th International Conference on LCA in Foods, April 25–26, 2007*, Gothenburg, Sweden.

Larsson, I. 2003. "Possible benefits from using LCA in the agrofood chain, example from Arla foods." In: *Life Cycle Assessment in the Agri-Food Sector: Proceedings from the 4th International Conference, October 6–8, 2003*, Bygholm, Denmark.

LCA Food. 2008. Available from http://www.lcafood.dk. Accessed August 28, 2008.

Madsen, J., and S. Effting. 2003. Using I/O data to find hot spots in LCA: Example of a hamburger meal. In: *Life Cycle Assessment in the Agri-food Sector: Proceedings from the 4th International Conference, October 608, 2003*, Bygholm, Denmark.

Nielsen P.H., Nielsen AM, Weidema BP, Dalgaard R and Halberg N. 2003. LCA food data base. www.lcafood.dk. Accessed August 28, 2008.

Ogino, A., H. Orito, K. Shimada, and H. Hirooka. 2007. Evaluating environmental impacts of the Japanese beef cow-calf system by the life cycle assessment method. *Anim. Sci. J.* 78(4):424–432.

Olesen, J.E. 2005. Drivhusgasser fra jordbruget—reduktionsmuligheder. *DJF rapport Markbrug* 113:161 (in Danish).

Pimentel, D., and M. Pimentel. 2003. Sustainability of meat-based and plant-based diets and the environment. *Am. J. Clin. Nutr.* 78:6608–6663.

SAIC. 2006. *Life Cycle Assessment: Principles and Practice*. Available from http://www.epa.gov/ORD/NRMRL/lcaccess/pdfs/600r06060.pdf. Accessed August 16, 2008.

Thomassen, M.A. 2008. *Environmental Impact of Dairy Cattle Production Systems—An Integral Assessment*, pp.1–113. Ph.D. thesis, Wageningen University, the Netherlands.

Thomassen, M.A., and I.J.M. de Boer. 2005. Evaluation of indicators to assess the environmental impact of dairy production systems. *Agric. Ecosyst. Environ.* 111 (1–4):185–199./

Thomassen, M.A., K.J. van Calker, M.C.J. Smits, G.L. Iepema, and I.J.M. de Boer. 2008. Life cycle assessment of milk production systems in the Netherlands. *Agric. Syst.* 96(1):95–107.

Thrane, M. 2003. *Environmental Impacts from Danish Fish Products*. Ph.D. dissertation, Ålborg University, Denmark, Department of Development and Planning.

Weidema, B. 2003. *Market Information in Life Cycle Assessments*. Technical report, Danish Environmental Protection Agency. Environmental Project No 863.

Weidema, B.P., B. Mortensen, P. Nielsen, and M. Hauschild. 1996. *Elements of an Impact Assessment of Wheat Production*, pp. 1–12. Institute for Product Development, Technical University of Denmark.

Weidema, B.P., A.M. Nielsen, H. Halberg, I.S. Kristensen, C.M. Jespersen, and L. Thodberg. 2005. Sammenligning af miljøpåvirkningen af konkurrerende jordbrugsprodukter. Miljørapport nr. 1028. Miljøstyrrelsen, Miljøministreriet.

Weidema, B.P., M. Wesnæs, J. Hermansen, T. Kristensen, and N. Halberg, 2007. *Environmental Improvement Potentials of Meat and Dairy Products*, p. 190. Final Report. Sevilla: Institute for Prospective Technology Studies.

Williams, A., E. Audsley, and D. Sandars. 2006. *Environmental Burdens of Agricultural and Horticultural Commodity Production—Defra-funded Projects IS0205 and IS0222*. Available from http://www.silsoe.cranfield.ac.uk/iwe/expertise/lca.htm. Accessed August 14, 2008.

Yakovleva, N. 2005. "Sustainability indicators for the UK food supply chain." In: *Measuring Sustainability of the Food Supply Chain Seminar, October 27, 2005*, Cardiff University.

Ziegler, F. 2003. "Environmental impact assessment of seafood products." In: *Environmentally-Friendly Food Processing*, eds B. Mattsson, and U. Sonesson. Cambridge, England: Woodhead Publishing Limited.

Chapter 6

Social Aspects of the Food Supply Chain

Kantha Shelke, Justin Van Wart, and Charles Francis

Food Safety, Health, and Nutrition

While a portion of the population, especially in the North, enjoys a comfortable standard of living and even suffers from overnutrition and obesity, close to one-fourth of humanity survives on less than $1 per day and sees little solution to their malnutrition dilemma, even in a world of apparent plenty. Approximately 18,000 children die each day from hunger (Lederer, 2007). Projections are that 1.2 billion people may be hungry by the year 2025 (Runge and Senauer, 2007). This widespread malnutrition (over- and undernutrition), however, is only one component of the health and safety concerns with the current food system.

Food Safety

In the United States, the Center for Disease Control and Prevention estimated that 76 million people are sickened, 325,000 are hospitalized, and 5,000 die annually from food poisonings (Mays, 1999). A global review of infectious disease found out food-borne disease ranks high in priority in the coming century (Cohen, 2000). The most prevalent food-borne pathogens are overwhelmingly associated with animal products, most of which come from factory farms and high-speed processing facilities. As a result, there is growing concern for practices employed by these operations such as feeding of antimicrobial agents to livestock as growth promoters, the feeding of rendered materials to food animals, increased emphasis on longer food shelf life and preservation by refrigeration only, high-speed operations, and the growing consolidation within the food industry (Cohen, 2000).

To begin with, the use of growth-promoting antibiotics in animal agriculture is one of the leading factors driving the increase in antibiotic resistance in humans (Horrigan et al., 2002; Sapkota et al., 2007). Antibiotic/antimicrobial resistance is a major public health crisis, eroding the discovery of antibiotics/antimicrobials and their application to clinical medicine. Although it is clear that the use of antibiotics in livestock has many proved *benefits*, it is equally clear that the use of these same antibiotics leads to "the increase of antibiotic-resistant bacteria of human significance (Mathew et al., 2007). Agricultural antibiotic use represents a major driver of antibiotic resistance worldwide for four reasons: (1) It is the largest use of antibiotics worldwide; (2) much of the use of antibiotics in agriculture results in subtherapeutic exposures of bacteria; (3) drugs of every important clinical class are utilized in agriculture; and (4) human populations are exposed to antibiotic-resistant pathogens via consumption of animal products as well as through widespread release into the environment. The firs reason related to the evolution and spread of microbial resistance, and is based on fundamental biology and evolution: the inevitable increase in antimicrobial resistance in response to exposure to antimicrobial agents. Over millennia, microbes evolved highly effective mechanisms to respond to external agents that threaten their lives (Wright, 2007). Exposure of bacteria to sublethal levels of antimicrobial agents is particularly effective in allowing the survival of resistant strains and, with continued antimicrobial pressure, resistant strains come to have an edge in terms of reproduction and spread. The rapidity of bacterial reproduction and replication allows for greater efficien y in the expression of these changes.

The second reason entails the fact that bacterial resistance to antimicrobials involves both genetic and regulatory changes. Regulatory changes, although rare, have more serious implications for the public health, because enhanced activity of physiological processes including membrane transport usually releases harmful agents, including antimicrobials. Genetically encoded changes are serious because they are persistent, confer high-level resistance to specifi or multiple agents, and can transfer among bacteria. Under selective pressure, bacterial populations quickly evolve to resistant phenotypes through mutation and sharing (Tenover, 2006).

That bacteria can share resistance genes is the basis of the third reason related to the evolution and spread of microbial resistance. Sharing genes that encode resistance is a robust survival mechanism by which resistance can be propagated within and among bacteria including commensals (nonpathogenic) and pathogens (Rowe-Magnus et al., 2002), often across broad species divisions. The latter facility is most dangerous in terms of propagation of resistance in the public health context (Summers, 2002) and accounts for more than 95% of antibiotic resistance (Nwosu, 2001), especially in resistant *Escherichia coli* isolated from consumer meat products (Sunde and Norstrom, 2006). This find ing is even more concerning because integrons can transfer multiple resistance genes at a time; some of these mechanisms can be enhanced by stressors, including antimicrobial pressure (Blázquez et al., 2002). Several studies attribute the environmental reservoirs of resistance for both poultry and swine to agricultural antimicrobials (Nandi et al., 2004; Jensen et al., 2006). Furthermore, the exchangeability of genes for resistance traits from commensal to specifi pathogens renders the formerly widely held *one bug, one drug* definitio totally inadequate (Summers, 2006).

The fourth and fina reason is that even after antimicrobials are no longer present, resistance may continue. Sørum et al. (2006) reported empirical evidence indicating that resistance decreased significant y even after the removal of antimicrobials from animal feeds, but could still be detected in animal houses, waste, and food products in Europe. Disposal of animal waste is a major route of environmental contamination by antimicrobials and resistance determinants. Farmers and farm workers, at significant y increased risks of infection by antimicrobial-resistant bacteria, may serve as entry points for the general community at large.

Emerging pathogens are also a growing concern, and when partnered with antibiotic resistance, the potential effects are alarming. This has been evidenced recently with the growing incidence of methicillin-resistant *Staphylococcus aureus* (MRSA). Another emerging concern about the food supply is a neurologic disease in cattle known as bovine spongiform encephalopathy that may cause degenerative neurologic disease in humans. This has also come about as a result of the factory farm practices described above.

Prevailing production methods have serious environmental and public health concerns. For example, monoculture agricultural practices erode biodiversity among plants as well as animals and require intensive chemical treatment. Horrigan et al. (2002) point out that the use of such synthetic chemical pesticides, herbicides, and fertilizers polluting soil, water, and air harms both the environment and the human health. Growth in animal products in the diet exacerbates these issues since the grain raised to supply feedlots (cattle) and factory farms (chickens, hogs, veal calves) is grown in intensive monocultures that require more chemical use (Horrigan et al., 2002).

The United States Environmental Protection Agency has found that current farming practices are responsible for 70% of the pollution in the nation's rivers and streams (Cook, 1998). Such pollution is known to raise risks for certain cancers as well as reproductive and endocrine system disorders (Horrigan et al., 2002). Herbicides, fungicides, and fumigants in every class of pesticide – organophosphates, carbamates, pyrethroids, and especially organochlorides – have at least one agent capable of affecting a reproductive or developmental endpoint in laboratory animals or people (Frazier, 2007). While many of the most toxic pesticides have been banned or restricted in developed nations, high exposures to these agents are still occurring in the most impoverished countries around the globe. Pesticides are linked with development abnormalities and deformities such as extra legs growing from the abdomens and backs in frogs (Ouellet et al., 1997). In humans, Frazier discovered that exposure of men or women to certain pesticides at sufficien doses may increase the risk for sperm abnormalities, decreased fertility in male children, spontaneous abortion, birth defects, or fetal growth retardation. Additionally, pesticides from workplace or environmental exposures enter breast milk (Jing et al., 2008; Mahjoubi-Samet et al., 2008; Polder et al., 2008; Raab et al., 2008; and Wang et al., 2008). Additionally, the two most commonly applied herbicides in the United States are suspected endocrine disruptors (Horrigan et al., 2002).

One concerning aspect is that the health effects of most pesticides, herbicides, and fertilizer are not known, especially with regard to their effects when used together. Further, exposure to these chemicals goes beyond contaminated water and includes residues on food and accumulation in foods. Since humans eat higher on the food chain (consuming animal products), this increases exposure to the chemicals as they accumulate at higher trophic levels, from plants, to animals, to humans (Horrigan et al., 2002).

Only recently transgenic hybrids (often called generically, genetically engineered foods) have been introduced into the human food supply.

One of the concerns surrounding these genetically engineered foods is that new allergens could be introduced into the food supply (USNRC, 2000). In addition, people with food allergies may have greater issues findin allergens in foods not previously a problem. Further, antibiotic resistant genes are used as markers in the genetic engineering of foods. This practice raises two possible concerns: eating such foods soon after taking antibiotics could reduce or eliminate the drugs' effectiveness because enzymes produced by the resistance genes can break down antibiotics; and resistance could be transferred to diseased organisms in the digestive tract, making it harder to treat them with antibiotics (Horrigan et al., 2002).

Health and Nutrition

Heller and Keoleian (2000) suggest, "A sustainable food system must be founded on a sustainable diet. In the most general sense, this would be a diet that matched energy intake with energy expenditure while supplying necessary nutrients for a healthy lifestyle." In 2000, the United States food supply provided 3,800 calories per person per day. Accounting for waste, the average person consumed 2,700 calories per day—an increase of 24.5% from 1970 (University of Michigan, 2006). This results in over 154 and 127% more caloric intake than recommended for men and women, respectively (Wilkinson, 1997). In 2004, 66% of United States adults were either overweight or obese, define as having a body mass index of 25 or more (University of Michigan, 2006). Diet contributes to heart disease, certain cancers, and stroke—the three major causes of United States deaths. The United States reliance on animal products in the diet, high in saturated fat, the only source of dietary cholesterol and low in fibe , is a leading cause for these chronic diseases (Horrigan et al., 2002). The average gets person in the United States 67% of their dietary protein from animal sources, compared to a 34% average worldwide (Pimentel and Pimentel, 1996). When countries shift to diets more like those in the United States, such as is being done in China where meat consumption nearly doubled countrywide during the 1990s, chronic diseases have become a more common cause of death (FAO).

In 2005, people in the United States ate 200 lb of meat per person, which is up 22 lb from 1970 (University of Michigan, 2006).

Recent estimates from public health experts suggest that a reduction of around 60% in daily intake of meat in developed countries would help reduce excess weight and obesity and offer other health benefit to individuals and society (Compassion in World Farming, 2008). This effect will also assist the environmental concerns associated with this type of agricultural production, given that animals inefficient y convert and use energy. For example, cattle are the most inefficien in their energy conversion, requiring 7 kg of grain to produce 1 kg of beef (compared to 4:1 for pork and 2:1 for chicken) (Brown et al., 1998). Further, over half of grains grown are fed to animals.

In addition to animal products, people in the United States are consuming more caloric sweeteners than ever before (Heller and Keoleian, 2000). Sugar has become the number one food additive, in the United States accounting for 16% of total caloric intake. Yet, corn sweeteners (especially high-fructose corn sweeteners) are rapidly replacing sucrose (made from cane and beets), due in large part to the consumption of carbonated nondiet soft drinks (Heller and Keoleian, 2000).

Food conservation, preservation techniques, and food processing, ensure we can eat at times of the year or in locations where fresh food is not readily available. There is an increase reliance on processed, convenience foods. Processing presents problems when it removes nutrients contained in the whole food products. For example, during flou production iron, niacin, thiamine, ribofl vin, and folate are replaced. Processed food also tends to be high in fat, salt, and sugar since this adds stability and decreases cost. In addition, there seems to be greater enjoyment of the taste and mouthfeel of foods high in salt, sugar, and fat, to add to our tendency to verconsume food.

Food additives are not a new phenomenon and some food additives have been around for centuries—for example, baking powder (bicarbonate of soda) has been used since the nineteenth century and vinegar (acetic acid) has been used since before Christian era to preserve foods and prevent microbial damage. While food additives are subject to rigorous safety testing before they are approved for use studies are increasingly questioning their safety and utility in food products.

Hyperactivity or attention-defici hyperactivity disorder (ADHD) is one of the most common neurological conditions found in children. Studies linking food additives to hyperactivity in children go as 30 years

(Feingold, 1975). Schab and Trinh's recent meta-analysis of double-blind, placebo-controlled trials provided evidence that artificia food additives and colors may have a significan effect on the behavior of children with ADHD (Schab and Trinh, 2004). The Food Standards Agency (FSA) in the United Kingdom commissioned a study that suggested that certain artificia food color additives used in foods and drinks are associated with an increase in hyperactive behavior in children (FSA, 2007). A recent study by McCann et al. (2007) found that certain mixtures of food additives including artificia colors and the preservative benzoate can lead to an increase in hyperactive behaviors including inattention, impulsiveness, and overactivity in some children.

Heart disease is a leading cause of death for adults in industrialized countries. Trans fatty acids, a product of fat/oil hydrogenation, have recently been linked to heart disease. Chemist Paul Sabatier discovered the process of hydrogenation using a nickel catalyst. Soon after, German chemist Wilhelm Normann used hydrogen gas to develop a commercial hydrogenation process and the process underwent several modification through the mid-twentieth century. Partially hydrogenated fat popularity accelerated in the 1960s, 1970s, and 1980s as the food industry's response to public health recommendations to reduce animal fats and tropical oils in the diet. Partially hydrogenated fats appeared as logical alternatives, particularly because of their stability, cost, availability, and functionality. Research studies conducted before the 1990s had limited data on the health effects of trans fats and the data often were contradictory. The variability was attributed to differences in the background diets, the fatty acid profil of the starting oil, differences in the extent and type of hydrogenation, and suboptimal comparison fats.

In 1990, Mensink and Katan studied the effect of diets rich in oleic acid, saturated fatty acids, and trans fatty acids and reported that cholesterol and low-density lipoprotein (LDL) cholesterol levels relative to an oleic acid–rich diet were higher after subjects consumed diets rich in saturated or trans fatty acids. In contrast to the trans fat–rich diet, high-density lipoprotein (HDL) cholesterol levels were higher after subjects consumed the diets rich in saturated fatty acids or oleic acid. When subjects consumed a trans fat–rich diet, the differential effects on LDL and HDL cholesterol levels resulted in the least favorable total-to-HDL-cholesterol ratio. Multiple research studies through the

1990s concurred the negative effect of partial hydrogenation and trans fats on serum cholesterol levels. Lichtenstein et al. (1999) reported that a linear positive relationship between partial hydrogenation and LDL cholesterol levels, increase in dietary partial hydrogenation level, increased the levels of LDL cholesterol. On the other hand, HDL cholesterol levels remained relatively constant with increasing hydrogenation and decreased only with the highest level of hydrogenation, creating the least favorable total-to-HDL-cholesterol ratio. The meta-analysis by Mozaffarian et al. (2006) found that a 2% increase of trans fats in the diet was associated with a 23% increase in the incidence of coronary heart disease. With this increasing evidence, the United States Food and Drug Administration (FDA) mandated the declaration of the trans fat content per serving in the Nutrition Facts panels of all packaged foods as of January 1, 2006. Food manufacturers, although previously aware of the research findings began to reformulate their products (reducing or removing trans fatty acid content) once the regulatory action was declared.

Consumer concerns over the quality and safety of foods, discussed above, are considered to be one of the primary reasons for the increasing demand for organically produced food, which consumers perceive as healthier and safer (Williamson, 2007).

Farmer and Market Effects

Ideally, the cost of goods and services would reflec ecological and social costs as well as what is typically included in budgeting the costs of production, processing, and marketing. In this context, competition and the law of demand would ensure that those goods produced and provided at "true" lower costs would flourish Capitalistic economics could work for us, but it often fails to achieve the most efficien production, consumption, and distribution of goods because of externalities and imperfect information. If we consider economics as one among several tools, rather than the *only* tool, clever solutions can be developed that will include price tags that reflec the true costs of products. An example of this is the recent prevalence of ecolableing in the United States, indicative of a needed shift in the market. Certificatio and labeling are ways to mitigate price misinformation and have become significan factors for both consumer and producer decisions. Labeling is an attempt to publicly verify adherence to including eco-costs or social costs into production practices and accounting

However, the current economic system still is set up to have very low prices for food, minimizing the ability for a small farm to compete due to their lower elasticity such as the need to farm more land to make the same amount of money. For example, the North American Free Trade Agreement trade community of Canada, United States, and Mexico—has been highly profita le for some large farmers and grain exporters in the United States, while at the same time has been disastrous for many small farmers in Mexico due to the drastically lower price for maize in the local marketplace. The current way of marketing and advertising results in short-term economic success, but in the food system this reflect neither nutritional value nor sustainability of production in an increasingly global food economy.

Small Farms and Rural Communities

It has been estimated that between 1910 and 1990, the share of the United States agricultural economy going to farmers declined from 41 to 9%, while the marketing and farm input industries' shares increased by similar amounts (Madden and Chaplowe, 1997). This reduction in profi for farmers has coincided with greater reliance on off-farm resources such as hybrid seed, pesticide, herbicide, and fertilizer purchases and greater government intervention with pricing and subsidies. The way to survive such a trend has been a shift to large-scale, monoculture farming. This shift has demonstrated challenges for the economic health of rural communities less jobs, less local retail spending, and lower local per capita income (Copeland and Zin, 1998). The loss of local food systems is also seen to have a spiritual dimension, since direct links between people and the farms that produce their food can enable people to reengage with the natural world, the local landscape, their local community, and the pleasures of eating fresh food in season.

Small farms generally are more diversifie than large, industrial farms, providing for more nutritionally dense foods and more environmentally preferable production methods. For example, small farms in the United States are estimated to have three times the trees per acre as larger farms and devote nearly twice as much of their land for cover crops and green manures (Rosset, 1999). It is also recognized that due to the diversificatio and land stewardship of small farms, they commonly produce more agricultural output per acre than larger farms (Baarret, 1993). So, if the market shifted to make small farms more feasible, then there will be social and environmental improvements.

The shift needed in the market is characterized by quality, rather than quantity. This will involve greater diversificatio in purchasing of agricultural products from industry (processors) and consumers. This will result in distribution moving from a national and global focus to a local and community-oriented approach. Across the United States we see evidence of this shift as more farmer markets and community organizations are marketing more directly to consumers.

While the government has and will continue to play a significan role in food marketing, ecolabels have been playing a much more signifi cant role since they can signal price and quality signals, and their sub- sequent opportunities and costs (e.g., Rainforest Alliance, Fairtrade). These standards and regulating abilities are expected to grow to meet public demand for quality and transparency.

Conclusions

Some of the leading social aspects of sustainability include, though are not limited to, the ability of the supply chain to provide nourishment to all of the population without undesirable health effects and the ability for diverse communities and farms to prosper. Consolidation of farms has led to a concentration of wealth, reductions in population, and loss of infrastructure in rural communities, as well as greater environmental degradation.

One key to providing nourishment globally is for the United States and other industrialized nations to lower their meat consumption. This would provide public health benefit in those countries, particularly a reduction in heart disease, cancer, other chronic diseases, and further increases in pathogenic illnesses. This potentially would also make the land used for feed grain production available for other crops or uses, providing more nutritionally dense foods and further alleviate health concerns associated with pesticide and herbicide use, often necessary for monoculture feed grain production.

To develop a truly sustainable agriculture and food system, it is essential that thoughtful people pursue a goal of equity to provide at least the minimum nutritional requirements needed by everyone worldwide. We should even seriously consider the elements of the charter of the United Nations from 1948, Article 25, where the identificatio of food as one of the basic human rights was specifie along with clean water, sanitation, freedom of assembly and movement, and many others. No solution short of this will provide a long-term, sustainable global society in which we would want to live. This is a large challenge to us and to the next generation.

References

Baarret, C.B. 1993. *On Price Risk and the Inverse Farm Size-Productivity Relationship.* University of Wisconsin-Madison, Department of Agricultural Economics Staff Paper Series No. 369.

Blázquez, J., A. Oliver, and J.-M. Gómez-Gómez. 2002. Mutation and evolution of antibiotic resistance: Antibiotics as promoters of antibiotic resistance? *Curr. Drug Targets* 3:345–349.

Brown, L.R., M. Renner, abd C. Flavin. 1998. *The Environmental Trends That Are Shaping Our Future.* New York: W.W. Norton.

Cohen, M.L. 2000. Changing patterns of infectious disease. *Nature* 406:762–767.

Compassion in World Farming. 2008. *Global Warning: Climate Change and Farm Animal Welfare.* Available from http://www.ciwf.org.uk/includes/documents/cm_docs/2008/g/global_warning.pdf. Accessed September 15, 2008

Cook, M. 1998. *Reducing Water Pollution from Animal Feeding Operations.* National Water Quality Inventory, EPA Report to Congress, EPA 841R001. Washington, DC.

Copeland, C., and J. Zin. 1998. *Animal Waste Management and the Environment: Background for Current Issues.* Report for Congress. Washington, DC: Congressional Research Service, FAO. FAOSTAT Database. Available from http://apps.fao.org/.

Feingold, B.F. 1975. Hyperkinesis and learning disabilities linked to artificia food fl vors and colors. *Am. J. Nurs.* 75:797–803.

Frazier, L. 2007. Reproductive disorders associated with pesticide exposure. *J. Agromedicine* 12(1):27–37.

FSA (Food Standards Agency). 2007. *Chronic and Acute Effects of Artificial Colourings and Preservatives on Children's Behaviour.* Available from http://www.food.gov.uk/multimedia/pdfs/additivesbehaviourfinrep.pdf Accessed September 15, 2008.

Heller, M., and G. Keoleian. 2000. *Life Cycle-Based Sustainability Indicators for Assessment of the U.S. Food System*. A report of the Center for Sustainable Systems University of Michigan. Report No. CSS00-04.

Horrigan, L., R.S. Lawrence, and P. Walker. 2002. How sustainable agriculture can address the environmental and human health harms of industrial agriculture. *Environ. Health Perspect*. 110(5):445–456.

Jensen, V.F., L. Jakobsen, H.-D. Emborg, A.M. Seyfarth, and A. Hammerum. 2006. Correlation between apramycin and gentamicin use in pigs and an increasing reservoir of gentamicin-resistant *Escherichia coli*. *J. Antimicrob. Chemother*. 58:101–107.

Jing T., Q.Q. Li, A. Loganath, Y.S. Chong, M. Xiao, and J.P. Obbard. 2008. Multivariate data analyses of persistent organic pollutants in maternal adipose tissue in Singapore. *Environ. Sci. Technol*. 42(7):2681–2687.

Lederer, E. 2007. U.N.: Hunger kills 18000 kids each day. *The Associated Press*, February 17, 2007.

Lichtenstein, A.H., L.M. Ausman, S.M. Jalbert, and E.J. Schaefer. 1999. Effects of different forms of dietary hydrogenated fats on serum lipoprotein cholesterol levels. *N. Engl. J. Med*. 340:1933–1940.

Madden, J.P., and S.G. Chaplowe. 1997. *For All Generations: Making World Agriculture More Sustainable*. Glendale, CA: World Sustainable Agriculture Association.

Mahjoubi-Samet, A., H. Fetoui, and N. Zeghal. 2008. Nephrotoxicity induced by dimethoate in adult rats and their suckling pups. *Pest. Biochem. Physiol*. 91(2):96–103.

Mathew, A., R. Cissell, and S. Liamthong. 2007. Antibiotic resistance in bacteria associated with food animals: A United States perspective of livestock production. *Foodborne Pathog. Dis*. 4(2):115–133.

Mays, J.J. 1999. CDC releases new estimates on foodborne diseases. *The Associated Press*, September 16, 1999.

McCann, D., A. Barrett, A. Cooper, D. Crumpler, L. Dalen, K. Grimshaw, E. Kitchin, K. Lok, L. Porteous, E. Prince, E. Sonuga-Barke, J.O. Warner, and J. Stevenson. 2007. Food additives and hyperactive behaviour in 3-year-old and 8/9-year-old children in the community: A randomised, double-blinded, placebo-controlled trial. *Lancet* 370:1560–1567.

Mensink, R.P., and M.B. Katan. 1990. Effect of dietary trans fatty acids on high-density and low-density lipoprotein cholesterol levels in healthy subjects. *N. Engl. J. Med*. 323(7):439–445.

Mozaffarian, D., M.B. Katan, A. Ascherio, M.J. Stampfer, W.C. Willett. 2006. Trans fatty acids and cardiovascular disease. *N. Engl. J. Med*. 354:1601–1613.

Nandi, S., J.J. Maurer, C. Hofacre, and A.O. Summers. 2004. Gram-positive bacteria are a major reservoir of Class 1 antibiotic resistance integrons in poultry litter. *Proc. Natl. Acad. Sci. U S A* 101:7118–7122.

Nwosu, V.C. 2001. Antibiotic resistance with particular reference to soil microorganisms. *Res. Microbiol*. 152:421–430.

Ouellet, M., J. Bonin, J.R. Rodrigue, J.-L. DesGranges, and S. Lair. 1997. Hind-limb deformities (ectromelia, ectrodactyly) in free living anurans from agricultural habitats. *J. Wildl. Dis*. 33:95–104.

Pimentel, D., and M. Pimentel, eds. 1996. *Food, Energy and Society*. Niwot, CO: University of Colorado Press.

Polder, A., G.W. Gabrielsen, J. Øyvind Odland, T. Savinova, A. Tkachev, K. B. Løken, and J.U. Skaare. 2008. Spatial and temporal changes of chlorinated pesticides, PCBs, dioxins (PCDDs/PCDFs), and brominated flam retardants in human breast milk from Northern Russia. *Sci. Total Environ.* 391(1):41–54.

Raab, U., U. Preiss, M. Albrecht, N. Shahin, H. Parlar, and H. Fromme. 2008. Concentrations of polybrominated diphenyl ethers, organochlorine compounds and nitro musks in mother's milk from Germany (Bavaria). *Chemosphere* 72(1): 87–94.

Rosset, P. 1999. *The Multiple Functions and Benefits of Small Farm Agriculture: In the Context of Global Trade Negotiations*. Oakland, CA: The Institute for Food and Development Policy. Available from http://www.foodfirst.o g/files/pb4.pdf Accessed September 15, 2008.

Rowe-Magnus, D., A.-M. Guerout, and D. Mazel. 2002. Bacterial resistance evolution by recruitment of superintegron gene cassettes. *Mol. Microbiol.* 43:1657–1669.

Runge, C.F., and B. Senauer. 2007. How biofuels would starve the poor. *Foreign Affairs* May/June 86(3).

Sapkota, A., F.C. Curriero, K.E. Gibson, and K.J. Schwab. 2007. Antibiotic-resistant enterococci and fecal indicators in surface water and groundwater impacted by a concentrated swine feeding operation. *Environ. Health Perspect.* 115(7):1040–1045.

Schab, D.W., and N.H. Trinh. 2004. Do artificia food colors promote hyperactivity in children with hyperactive syndromes? A meta-analysis of double-blind placebo-controlled trials. *J. Dev. Behav. Pediatr.* 25:423–434.

Sørum, M., P.J. Johnsen, B. Aasnes, T. Rosvoll, H. Kruse, A. Sundsfjord, and G.S. Simonsen. 2006. Prevalence, persistence, and molecular characterization of glycopeptide-resistant enterococci in Norwegian poultry and poultry farmers 3 to 8 years after the ban on avoparcin. *Appl. Environ. Microbiol.* 72:516–521.

Summers, A.O. 2002. Generally overlooked fundamentals of bacterial genetics and ecology. *Clin. Infect. Dis.* 34(3):S85–92.

Summers, A. 2006. Genetic linkage and horizontal gene transfer, the roots of the antibiotic multi-resistance problem. *Anim. Biotechnol.* 17:125–135.

Sunde, M., and M. Norstrom. 2006. The prevalence of associations between and conjugal transfer of antibiotic resistance genes in *Escherichia coli* isolated from Norwegian meat and meat products. *J. Antimicrob. Chemother.* 58:741–747.

Tenover, F.C. 2006. Mechanisms of antimicrobial resistance in bacteria. *Am. J. Med.* 119:S3–S10; discussion S62–S70.

University of Michigan. 2006. *U.S. Food System Fact Sheet*. Available from http://css.snre.umich.edu/css_doc/CSS01-06.pdf//. Accessed September 15, 2008.

USNRC (United States National Research Council). 2000. *Genetically Modified Pest-Protected Plants: Science and Regulation*. Washington, DC: National Academy Press.

Wang, Y.-R., M. Zhang, Q. Wang, D.-Y. Yang, C.-L. Li, J. Liu, J.-G. Li, H. Li, and X.-Y. Yang. 2008. Exposure of mother–child and postpartum woman–infant pairs to DDT and its metabolites in Tianjin, China. *Sci. Total Environ.* 396(1):34–41.

Wilkinson, C.W., J.D. Goldman, and A. Cook. 1997. Trends in food and nutrient intakes by adults. *Fam. Econ. Nutri. Rev.* 10(4):2–15.

Williamson, C.S. 2007. Is organic food better for our health? *Nutr. Bull.* 32(2):104–108.

Wright, G.D. 2007. The antibiotic resistance: The nexus of chemical and genetic diversity. *Natl. Rev. Microbiol.* 5:175–186.

Chapter 7

Ecolabeling and Consumer Interest in Sustainable Products

Amarjit Sahota, Barbara Haumann, Holly Givens,
and Cheryl Baldwin

Introduction

Environmental and social issues are moving up in food purchasing priorities (Sloan, 2007). Approximately 50% of the United States consumers consider at least one sustainability factor in selecting consumer packaged goods items and choosing where to shop for those products, according to a survey conducted by Information Resources, Inc. (IRI, 2007). As mentioned in the Introduction to this book, purchases of sustainable foods have been led by consumer's interest in their health. Major considerations have been on products free of pesticides, hormones, antibiotics, and genetically modified ingredients (Molyneaux, 2007). However, there is a growing shift in priorities. Attributes concerning social issues are becoming more important than other priorities. In the BBMG Conscious Consumer Report (BBMG, 2007), close to 90% of Americans identified well with *conscious consumers*, *socially responsible*, and *environmentally friendly*. BBMG (2007) found that convenience was lower in purchase priority than attributes like where a product is made or how energy efficient it is.

In addition to the food's attributes, consumers are growing more aware and interested in the companies they are buying food from, and state that they prefer to purchase from companies that support social, community, and environmental interests (Molyneaux, 2007).

Consumer Insight and Green Marketing

Each consumer segment has some level of interest in more responsible (socially and environmentally) products. Market research consistently shows that the most dedicated group of socially conscious purchasers is about 10% of the population. This group is well informed and driven significantly by values. However, the majority of purchasers comprise a more moderate perspective and aim to balance convenience, cost, and making a difference. For this, and the very motivated group of purchasers, trust in the company and the claims/marketing is critical. Overpromising or misrepresenting benefits, called *greenwashing*, is eventually uncovered.

There is a portion of the population with very little motivation to purchase socially responsible products. However, they will make such purchases if they meet other values (quality, price, and convenience), and they can see how their contribution helps make a difference. In general, all consumers want to make a contribution and be rewarded for their commitment. As a result, the most effective sustainability-related messages are personalized and credible.

Typically, environmental claims on products follow an evolution, beginning with basic product characteristics such as recyclable package, then to differentiated performance such as life cycle–based claims, and finally claims to reflect how responsible and committed the company is, such as a claim about renewable energy. While the last means of communication is a more holistic message, a company making any type of environmental claim should practice responsibly to be credible and to gain long-term success. This commitment builds trust from consumers, and consumers reward such companies. Some of the desired company practices are production of energy-efficient products, products with health and safety benefits, support for fair labor and trade, and commitment to environmentally responsible practices (BBMG, 2007).

In the United States, the Federal Trade Commission (FTC) enforces the FTC Act to ensure that marketing is not deceptive or unfair, looking at marketing from a reasonable consumer's point of view. As a result, it is important to know the consumer since there is confusion about what many sustainability-related messages mean. A consumer may assume all plastic packaging is recyclable—though only limited types are actually accepted for recycling.

The FTC has developed Green Marketing Guides to address typical sustainability-related claims. In general, their guidance is that the message be appropriate, accurate, verified, and informative. Some key questions to consider when developing a message are as follows:

- What is expressed and implied?
- What do I need to substantiate the claim?
- Is qualification of the claim needed?

The Role of Certification

Some examples of foods sold at retail considered to represent at least some aspect of sustainable production include organic, fair trade, locally produced, no animal products, recyclable, no genetically modified organisms, and vegetarian. A growing category is certified products. Certified products serve as one of the top sources of information on the sustainability of a product or company (BBMG, 2007). Certification means that the product has been verified to meet a set standard, as evaluated by a third party. Certification also includes communication about this verification on the label, with a seal or mark. This means of delivering credible and validated information is important since the label continues to be one of the top sources of information for consumers when making purchasing decisions. Ecolabels, certification to standards with environmental requirements, are a way of recognizing products that have some attributes of sustainability.

A United States organization, Consumers Union, launched a web site (http://www.ecolabels.org) in 2001 to help inform consumers of the credibility of ecolabels and what they mean. The Consumers Union determines credibility based on the meaning/standards of the label, if an organization verifies that the label standards are met, the consistency of the meaning of the label, if the label standards are publicly available, if information about the organization is publicly available, if the organization behind the label is free from conflict of interest, and if the label was developed with broad stakeholder input (Consumers Union, 2002). The United Kingdom government has a list of ecolabels available to its population (DEFRA, 2007) and Ilbery and Maye (2007) summarized many of the ecolabels seen in the European Union (EU).

Organic Food

"Organic" is the dominant ecolabel for food (the details about the standards and programs that define and enforce 'organic' are covered later in this chapter). Organic Monitor estimates that global sales of organic food and drink surpassed US$40 billion in 2007 (Organic Monitor, 2008a). In comparison, the global market for other sustainable foods is estimated to be worth less than US$20 billion. Demand for organic products is concentrated in regions where there is high consumer awareness of ethical issues and where consumers have the purchasing power to support their beliefs.

In general, organic foods are grown without the use of toxic and persistent pesticides, synthetic fertilizers, and growth hormones. They are produced according to production standards that prohibit the use of genetically modified organisms, radiation, and sewage sludge. Organic livestock standards require living conditions appropriate to the species, including access to the outdoors and conditions that allow for the natural behavior needs of the animals, and prevent the routine use of antibiotics and growth promoters.

Organic sales in the United States are expected to double between 2007 and 2011, reaching US$35 billion and around US$65 billion globally (Molyneaux, 2007). Much of this is due to the regulations that have developed for organic certification (Molyneaux, 2007). Organic agriculture regulations have been set in the United States and many other countries around the world including the European Union (27 countries), Croatia, Iceland, Macedonia, Montenegro, Norway, Switzerland, Turkey, China, Japan, South Korea, Taiwan, Thailand, Argentina, Brazil, Chile, Ecuador, Honduras, and Tunisia. In Canada, which has had a strong voluntary organic standard since 1999, Organic Products Regulations will be fully implemented by June 30, 2009, and will require mandatory compliance for products marketed as organic (Huber et al., 2007).

Market and Consumer Interest in Organic Foods

Organic foods are one of the fast-growing market segments within the food industry, with sales growing at an annual rate of 21% in 2006 in the United States (Organic Trade Association, 2007). Consumer demand for organic products is concentrated in North America and Europe; these two regions comprise 97% of global revenues. Other regions like Asia, Latin America, and Australasia are important producers and exporters of organic foods; however, their internal markets are just beginning to develop.

Europe

Europe is one of the largest and most developed markets for organic food and drink in the world, valued at about US$20 billion in 2006. Organic food sales are concentrated in Western Europe, with Central and Eastern Europe having less than 3% share.

The four largest consumer markets—Germany, France, Italy, and the United Kingdom—comprise over 75% of regional revenues. Other countries like Denmark, Sweden, and Austria are showing high growth; however, they have much smaller markets because of their small consumer numbers.

The German and the United Kingdom markets continue to show the fastest growth in Europe. High growth in the German market is being driven by the entry of the major retailers.

Organic foods are becoming widely available in supermarkets, discount stores, department stores, and drugstores. Discounters like Aldi, Lidl, and Penny have had a major impact as they have made organic foods available to consumers at low prices. Supply shortages in both the German and the United Kingdom markets are stunting market growth rates, with retailers and food processors unable to find adequate supply of organic products.

Scandinavian and Alpine countries have the highest market share of organic food sales (Organic Monitor, 2008a). In countries like Denmark, Sweden, and Switzerland, organic food sales represent about 5% of total food sales (Organic Monitor, 2008a). For products like organic milk and eggs, the market share is as high as 30% in some countries (Organic Monitor, 2008a). The Swiss spend the most on organic food, with average spending of over US$150 per capita (Organic Monitor, 2008a). In contrast, Southern, Central, and Eastern European consumers spend the least on organic foods (Organic Monitor, 2008a).

A study by Dialego found that consumers generally have a positive attitude to the price premium of organic foods (Organic Monitor, 2008b). Two-thirds of consumers stated that it is right that organic food is more expensive, while only 10% stated the higher prices are not justified (Organic Monitor, 2008b).

More than half of British consumers buy organic fruit and vegetables (Organic Monitor, 2008b). About half of all consumers who claim to buy organic products believe they are too expensive and that they would buy more if they were cheaper (Organic Monitor, 2008b). However, research by Sainsbury's discovered that organic foods were no longer bought by just affluent consumers, with 31% of lower income households buying these products (Organic Monitor, 2008a). Further, equally high levels of men and women were buying organic foods, 30 and 38%, respectively.

A survey by Agence Bio found that 81% of French men and women had a positive impression of organic products (Organic Monitor, 2008a). When questioned on their reasons for buying organic foods, 94% stated *to stay healthy*, while 93% stated *to be certain of buying a healthy product* (Organic Monitor, 2008a). Other important factors are *quality and taste*, quoted by 90% of consumers (Organic Monitor, 2008a). A survey in Italy found that the main reason consumers buy organic products is the absence of pesticides and chemical fertilizers, stated by 86–96% of respondents (Organic Monitor, 2008a). The absence of genetically modified organisms is an important reason for 75–91% of organic food buyers (Organic Monitor, 2008a). Other important reasons are environmental protection (69–83%) and food safety (59–78%) (Organic Monitor, 2008a). Studies in other countries (Ireland, Spain, and Austria) had similar findings (Organic Monitor, 2008a).

North America

Organic food and drink sales in North America continue to surge, with the United States retail sales reaching US$16.7 billion in 2006, and projected to grow to US$23.6 billion in 2008 (Organic Trade Association, 2007). The United States has the largest market for organic products in the world. Significant increases in organic farmland are also making it a leading producer; almost all types of organic crops are grown in the United States; however, imports are necessary because of demand exceeding supply. Latin American countries like Mexico, Argentina, and Brazil export significant quantities of organic foods to the United States market.

Organic foods are now present in every food retail channel, whether it be in restaurants offering organic beef or mass merchandisers offering hundreds of stock-keeping units of organic products. As a result, more and more United States consumers, regardless of income, have access to organic foods because these foods are now in the same stores where they already shop.

In fact, mass market grocery stores represent the largest single channel for organic products, accounting for 38% of the United States organic food sales in 2006. In addition, another 8% of organic food and beverage sales were through mass merchandisers and club stores. Sales have also increased in food service outlets as well as via internet and mail order sales (Organic Trade Association, 2007).

The fastest-growing food categories in 2006 were organic meat sales (29% growth), dairy product sales (25% growth), fruit and vegetable sales (24% growth), and bread and grain sales (23% growth). The fastest-growing nonfood categories in 2006 were pet food sales (37% growth), household product and cleaner sales (31.6%), fiber sales (27% growth), and supplement sales (26% growth) (Organic Trade Association, 2007).

The introduction of Canadian national regulations for organic agriculture and organic foods is expected to give the market a boost. The U.S. market has been reporting high growth since the United States Department of Agriculture (USDA) implemented the National Organic Program (NOP) in 2002. Another major driver of market growth in both countries is increasing distribution in conventional grocery channels; large retailers like Wal-Mart, Safeway, and Loblaws are expanding their organic product ranges, with many introducing private-label organic products. The success of the Safeway 'O Organics' private label has led it to start exporting these organic products to the Asian market. Safeway 'O Organics' is already a leading brand of organic foods in North America; it is poised to become a global organic brand.

Within the United States, some major cities have a higher proportion of consumers who report using organic products than others. Scarborough Research (2007) found that 35% of San Francisco adults bought organic foods during September 2007. The same month, 32% of Seattle adults consumed organic foods, making it the second leading American city. Other leading cities are Portland (27%), Washington, DC and Denver (26% each), and San Diego (24%). In the United States overall, 17% of all adults are considered to be core organic food consumers.

Scarborough Research (2007) found that organic food consumers spend on average US$127 on their weekly household grocery bill, 10%

higher than the national average of US$115. The organic food consumer has a relatively high disposable income, with annual household income of US$86,000 a year, 22% higher than the national average. Scarborough Research also revealed that organic food consumers tend to be young with families. They are 19% more likely than the national average to be aged between 18 and 34, and 13% more likely to have two or more children at home.

A study by Mambo Sprouts Marketing in August 2007 found that most American consumers do not have a good understanding of the USDA Organic seal (Organic Monitor, 2008a). A poll of 850 natural and organic food consumers found that just under half of respondents thought the USDA organic seal indicated 100% organic contents (Organic Monitor, 2008a). A quarter of consumers thought the seal meant at least 95% organic, while 16% thought it indicated a product was more than 70% organic, and 12% felt it meant "some organic" (Organic Monitor, 2008a). The survey also revealed that consumers preferred foods that were both local and organic. When asked to choose between the two, 36% of consumers said they would choose local products over organic items, while 33% indicated the opposite (Organic Monitor, 2008a). The remaining respondents said they were unsure which to choose (Organic Monitor, 2008a).

A half of all American adults sometimes buy organic foods, according to the 2007 Harris Interactive survey (Organic Monitor, 2008a). The research company found that the reasons for buying organic foods varied. The main reason given by 40% of consumers is health benefits such as the fact that organic foods are produced without the use of pesticides and preservatives. Smaller numbers cite things such as organic food tastes better (8%), it is more environmentally responsible to buy organic food (8%), organic foods are fresher (6%), it is more socially responsible to buy organic foods (4%), and organic foods are better for my children (3%). Thus, avoidance of pesticides and hormones are reasons behind these products being entry points. Price was cited as the main reason why consumers do not buy more organic foods.

Asia

Asia is becoming an important producer and consumer of organic foods. China, with the largest organic farmland area in the region, is well established as a global source of organic seeds, beans, herbs, and ingredients.

India, Thailand, and the Philippines are also becoming important producers and exporters.

Retail sales of organic food were about US$780 million in 2006 (Organic Monitor, 2008a). Demand is concentrated in the most affluent countries, namely, Japan, South Korea, Singapore, Taiwan, and Hong Kong. As in other parts of the world, demand is surpassing supply with large volumes of organic foods imported into each country.

The Asian market is showing high growth because of widening availability and rising consumer awareness. A growing number of conventional food retailers are introducing organic products. The number of dedicated organic food shops is also rising, with many new store openings in countries like Singapore, Malaysia, and Taiwan. Some large food companies are also coming into the market and introducing organic lines.

Consumer awareness of organic foods is rising partly because of the high incidence of health scares in recent years. The scares, some involving foods, are raising consumer awareness of health issues and stimulating consumer demand for organic products. Health scares include avian flu and severe acute respiratory syndrome (SARS) and those involving foods included cola drinks (India) and tofu (Indonesia).

Although organic food sales are rising, consumer demand remains subdued partly because of the low spending power of most Asian consumers. Organic food prices are exceptionally high in some Asian countries. In Japan, Taiwan, and Singapore, some organic foods are priced four to five times as much as nonorganic foods. Since most finished organic products come in from countries like Australia and the United States, distribution costs and import tariffs inflate product prices.

Oceania

Australasia is an important producer of organic foods, though it is not yet a large consumer. Valued at about US$340 million, the Australasian market for organic food and drink comprises less than 1% of global sales (Organic Monitor, 2008a). Small consumer markets and the export focus of producers are responsible for the small market size.

Australia and New Zealand are international exporters of organic products. Leading exports include organic beef, lamb, wool, kiwi fruit, wine, apples, and pears. Although exports continue to increase, the portion of exports to total production is in decline as internal markets for organic food and drink develop.

The Australasian market for organic products is growing at a steady rate. Most sales are of organic fresh products like fruit, vegetables, milk, and beef; however, there is a rise in organic food processing. The number of mainstream food retailers selling organic products is increasing, while new organic food shops continue to open.

Other Regions
Production and sales of organic products are also increasing in other regions like Latin America, the Middle East, and Africa. Organic crops are grown across the South American subcontinent. Countries like Argentina, Brazil, and Chile have become important producers; however, over 90% of their organic crops are destined for export markets. Most organic food sales in these countries are in major cities like Buenos Aires and São Paulo. Organic food production in Africa is almost entirely for the export market. Middle Eastern cities like Dubai, Riyadh, and Kuwait City, however, are becoming important consumers of organic products.

Global Organic Standards and Regulations

The International Organic Federation of Agriculture Movements (IFOAM) published the first international organic standards in 1980. IFOAM (http://www.ifoam.org), a worldwide umbrella organization for the organic movement, includes more than 750 member organizations in 108 countries. IFOAM has an organic guarantee system, providing a market guarantee for the integrity of organic claims through a common system of standards, verification, and marketing identity. It also fosters equivalence among participating IFOAM-accredited certifiers.

Several European countries, such as Austria and France, adopted legislative measures in the 1980s concerning organic agriculture. In 1991, the European Union adopted its organic regulation 2092/91, which included standards for organic production, labeling, and inspection. This regulation, which went into effect in January 1993, underwent a revision process beginning in 2005, with the European Union Agricultural Council reaching agreement in December 2006 on the general direction of the new proposed regulation, slated to go into effect January 1, 2009 (Huber et al., 2007).

Meanwhile, the Codex Alimentarius Commission, a joint Food and Agriculture Organization/World Health Organization (FAO/WHO) committee, over time has developed guidelines concerning aspects of organic production, processing, marketing, and labeling (Huber et al., 2007).

The National Organic Program in the United States, administered by the United States Department of Agriculture, is considered to have the strictest requirements for organic certification in the world.

In places without government regulation of organic production, private standards may be in use. For instance, organic food producers in countries like New Zealand adopted private standards. Private standards are also in use in Europe and America for products that are outside of the scope of current regulations, such as processing of textiles made with organic fiber. The growth of the organic food industry has led to a plethora of certification agencies. In 2007, 468 organizations were involved in certifying organic products, up from 383 in 2004 (Willer et al., 2008). Differences between organic standards can sometimes hamper global trade of organic products. For instance, there is little intra trade of organic processed foods between the European Union and the United States because of standard differences (Organic Monitor, 2008a).

The main markets for organic producers are the European Union, Japan, and the United States. A booklet entitled "Certification of Organic Foodstuffs in Developing Countries" describes the framework for the import of organic products from developing countries to the European Union, Japan and, the United States (Neuendorff).

European Organic Standards and Certification
The major European Community Regulation applying to organic food is the Council Regulation (EEC) No. 2092/91 of June 24, 1991. It covers organic production of agricultural products and foodstuffs, according to the summary of the Official Journal Reference L198 of July 22, 1991:

> The regulation establishes a common framework of minimum statutory standards throughout the community for agricultural foodstuffs bearing, or intended to bear, indications referring to organic production methods. It lays down rules on the production, inspection, processing and labeling of all such foodstuffs marketed in the community including imports.

The purpose of the European regulations is to provide a harmonized framework for the labeling, production, and inspection of organic foods throughout the European Union.

Article 11 of the Council Regulation (EEC) No. 2092/91 ensures imported organic foods follow the same strict criteria. The summary:

> Article 11 ... provides for a control regime on products imported from third world countries. From 1 January 1993 such products may only be imported from third world countries appearing in a list to be drawn up in accordance with regulatory committee procedure. To appear in the list the third country's public authority must apply for recognition of equivalence of the arrangements applied to its territory.

Imported products must meet European Union organic regulations to be labeled and marketed as organic. The organic standards of seven countries are recognized and approved by the European Union. The organic products of this "third country list" therefore do not need to be recertified for the European market. Countries on this list are Argentina, Australia, Costa Rica, Israel, New Zealand, Switzerland, and India.

Organic products can also come into the European Union via equivalency arrangements between certification agencies. If the organic standards of the foreign certifier have equivalency with standards of a European certifier, then the organic fruit and vegetables do not need to be reinspected and recertified. For instance, the Soil Association (United Kingdom) and California Certified Organic Farmers (United States) have had an equivalency arrangement since 2003. This allows American products certified by California Certified Organic Farmers to have the Soil Association certification for United Kingdom sales. Other organic products can be imported on a case-to-case basis by European Union member countries.

The European Union passed a new regulation for organic foods in June 2007. Regulation (EC) No. 834/2007 is for organic production and labeling of organic products and repeals Regulation (EEC) No. 2092/91. The new regulation sets out a complete set of objectives, principles, and basic rules for organic production. It also includes a new import regime and a more consistent control regime. Regulation (EC) No. 834/2007 will come into force on January 1, 2009.

Although many European countries have national regulations for organic agriculture and organic foods, Denmark is the only country

Figure 7.1.   Some of the organic logos and symbols in Europe.

with state inspection and certification services. Organic products that meet Danish state regulations are given the government logo, the "Ø" mark (Ø for økologisk). Norway and Switzerland are not members of the EU and both countries have national standards. Norway's standards are based on European Union regulations, and Switzerland has more stringent national standards. Figure 7.1 shows some of the organic logos and symbols in Europe.

Europe has over 160 certification agencies; however, there are no regional organizations that represent the European organic food industry. The most important is IFOAM. As mentioned, IFOAM has developed standards for organic farming and organic food production, which are designed to harmonize standards worldwide. Certification agencies must use IFOAM-equivalent standards in order to get IFOAM accreditation. Some European retailers prefer to import organic products that are certified by an IFOAM-accredited organization. Examples are Sainsbury's in the United Kingdom and Gröna Konsum in Sweden. In January 2008, 36 certification agencies were IFOAM accredited.

United States Organic Standards and Certification
In 1990, the United States Congress passed the Organic Foods Production Act (OFPA) as part of the 1990 Farm Bill. Its purpose was to establish national standards for the production and handling of foods labeled

as organic in the United States. Although private and state agencies had certified organic practices, there had been no national requirement for certification and thus no guarantee that organic meant the same thing from state to state, or even locally from certifier to certifier. Both consumers and producers of organic products sought national standards to clear up this confusion in the marketplace and to protect against mislabeling or fraud.

OFPA authorized the formation of the National Organic Program to establish organic standards and to require and oversee mandatory certification of organic production. In late December 2000, the USDA, which oversees the program, published the national organic standards final rule. These standards became fully implemented in October 2002.

OFPA also created the National Organic Standards Board (NOSB) to advise the Secretary of Agriculture in setting the standards on which USDA's National Organic Program is based. After considering the recommendations of NOSB, the Secretary has final authority in determining the regulations. Among its tasks, NOSB is authorized to convene technical advisory panels to advise which materials are included on a national list of materials, allowed or prohibited for use in organic production. The regulations were set up to evolve as the industry grows. For example, there are sunset provisions to reexamine materials allowed and prohibited in organic production, so that as more environmentally sound materials become available, the use of less environmentally sound materials can be phased out.

Under national organic standards, certification agencies must be accredited by USDA in order to certify producer, handling, or processing operations as organic. USDA posts a list of all accredited certifying agencies, including those based in the United States and those throughout the world. As of January 2008, there were 55 domestic certification agencies and 40 agencies outside the United States that were accredited as meeting the certifier requirements of the National Organic Program (USDA).

As outlined in the national organic standards, organic products are produced without the use of toxic and persistent pesticides, genetic engineering, irradiation, sewage sludge, synthetic growth hormones, or antibiotics. Although the first draft of the standards asked if the use of genetically modified organisms, sewage sludge, and irradiation were compatible with organic production, comments from over a quarter

Figure 7.2. USDA NOP logo.

million people kept these materials from being allowed for use in organic production.

In order to be certified organic, land must be free from the use of prohibited materials and practices for at least 3 years. In addition, cloning of animals is not allowed in organic production. Consumers seeking to avoid buying products derived from use of genetic engineering, sewage sludge, irradiation, animal cloning, and other prohibited methods can look for the organic label for such assurance.

All organic products, whether made in the United States or imported, must meet these national standards if they are to be marketed as organic in the United States. The program has been successful in that it has harmonized organic standards and it has produced a common organic logo (Figure 7.2). Thus, the national organic standards offer the United States consumers the assurance that all food products labeled as organic in the United States are governed by consistent standards (http://www.ams.usda.gov/nop/Consumers/Seal.html).

The United States organic standards allow four different labeling options based on the percentage of organic ingredients in a product. These include three distinct categories, and a fourth option for products that contain organic ingredients but not at a high enough level to meet one of the three labeling categories:

- *Category 1: 100% organic.* Only products that have been exclusively produced using organic methods are allowed to carry a label declaring "100% organic."
- *Category 2: Organic.* This signifies that at least 95% of the ingredients (by weight, excluding water and salt) in a processed product have been organically produced. The remaining contents can only be natural or synthetic ingredients recommended by the National Organic Standards Board and allowed on the National List.

- *Category 3: Made with organic.* Products with 70–95% organic ingredients may display "Made with organic [with up to three specific organic ingredients or food groups listed]" on the front panel.
- Products with less than 70% organic ingredients can list the organic items only in the ingredient panel. There can be no mention of organic on the main panel.

In the first two labeling categories (100% organic and organic), the product cannot use both organic and nonorganic versions of any ingredient that is listed as organic. For instance, if a loaf of organic bread is made with wheat, all of the wheat in the bread must be organic. All three categories prohibit the inclusion of any ingredients produced using genetic engineering, irradiation, sewage sludge, or cloning.

To assist consumers, the USDA has designed a seal that may be used only on products labeled as "100% organic" or "organic." Use of the seal is voluntary, but is seen as a useful tool to market to consumers.

The actual percent of organic content may be displayed on all products, regardless of label category. However, the rule specifies the actual dimensions that are allowed in displaying the content, and the percentage for products with less than 70% organic ingredients can only be displayed in the information panel.

Fairtrade Foods

Fairtrade is a trading partnership based on dialog, transparency, and respect that seeks greater equity in international trade. It contributes to sustainable development by offering better trading conditions to, and securing the rights of, marginalized producers and workers. Fairtrade labeling is undertaken by a network of independent, nonprofit national organizations. These organizations or labeling initiatives are present in 20 countries.

The first Fairtrade label was applied to coffee in the Netherlands in 1988. Since then, Fairtrade standards have been set for a range of products from the developing world including tea, sugar, cocoa, fresh fruits and vegetables, honey, wine, nuts, and spices. Nonfood products include flowers, plants, sports balls, and seed cotton.

Stichting Max Havelaar initiated fairtrade by giving products independent certification and enabled consumers to track the product origins.

Figure 7.3. Fairtrade logo.

Similar fairtrade labeling initiatives were set up in Europe and North America in the 1990s. These organizations came under the umbrella organization Fairtrade Labelling Organisation International (FLO) in 1997. FLO introduced the international Fairtrade mark in 2002. This is now internationally recognized as the Fairtrade symbol (Figure 7.3).

Market and Consumer Interest in Fairtrade

The global market for Fairtrade products was worth about US$2.2 billion in 2006 (Organic Monitor, 2008b). The market has more than doubled from US$1 billion in 2004. Like other markets for sustainable and ethical products, demand is concentrated in Europe and North America. The two regions comprise 99% of global revenues. The largest markets for Fairtrade products are in the United States (US$684 million), the United Kingdom (US$561 million), France (US$219 million), and Switzerland (US$185 million). Table 7.1 lists the leading Fairtrade product markets.

Production of Fairtrade products is concentrated in the developing world, whereas demand is almost entirely in the developed world. This large disparity between production and demand makes the market highly dependent on international trade, and country markets cannot become self-sufficient.

Fairtrade products are produced in about 60 countries by almost 1 million producers. Most are located in Latin American countries like Colombia, Mexico, and Bolivia and African countries like Kenya, South Africa, and Ghana. However, production is also increasing in Asian countries like India, Thailand, and Indonesia. Table 7.2 lists the leading Fairtrade products in the world market, showing that food and drink

Table 7.1. Leading markets for Fairtrade products (2006): Estimated retail sales (million US dollar)

United States	684
United Kingdom	561
France	219
Switzerland	185
Germany	151
Canada	74
Austria	57
The Netherlands	56
Italy	47
Belgium	38
Finland	31
Denmark	29
Sweden	22
Ireland	16
Norway	12
Australia and New Zealand	10
Japan	6
Luxemburg	4
Spain	3
Total	2,204

Note: All figures are rounded.
Source: Fairtrade Labelling Organizations International.

Table 7.2. Leading Fairtrade products (2006)

	Volume	Growth (2005–2006)
Bananas (tonnes)	135,763	31%
Cocoa (tonnes)	10,952	94%
Coffee (tonnes)	52,077	53%
Cotton (items)	1,551,807	125%
Flowers (stems)	171,697	51%
Honey (tonnes)	1,552	17%
Juices (tonnes)	7,065	45%
Rice (tonnes)	2,985	75%
Sports balls (items)	56,479	138%
Sugar (tonnes)	7,159	98%
Tea (tonnes)	3,886	49%
Wine (liters)	3,197,410	139%

Note: All figures are rounded.
Source: Fairtrade Labelling Organizations International.

products, especially commodities, are the dominant Fairtrade products (Fairtrade Labelling Organisation, 2007).

Fairtrade Standards and Certification

The Fairtrade mark is a certification label awarded to products sourced from the developing world that meet internationally recognized standards of fair trade. The Fairtrade mark provides five guarantees:

1. Farmers are given a fair and stable price for their products.
2. Farmers and estate workers are given extra income to improve their lives.
3. Production involves a greater respect for the environment.
4. Small farmers get a stronger position in world markets.
5. A closer link is developed between consumers and producers.

There are two sets of Fairtrade producer standards: one for small farmers and one for workers on plantations and in processing factories. The first set applies to smallholders organized in cooperatives or other organizations with a democratic, participative structure. The second set applies to organized workers, whose employers pay decent wages, guarantee the right to join trade unions, and provide decent housing, where relevant. On plantations and in factories, minimum health and safety as well as environmental standards must be complied with, and no child or forced labor can occur.

Agricultural products that are made according to these standards are certified Fairtrade. Products that contain certified Fairtrade and ingredients that are outside this system are referred to as composite products. Such products are allowed to have the Fairtrade mark if they contain at least 20% certified Fairtrade ingredients by weight. This allows products like juices, yoghurts, and snack bars to be certified Fairtrade.

FLO is responsible for setting and maintaining the standards that apply to producers and trading relationships. Through its certification company (FLO-Cert), it inspects and certifies producers against the standards, and audits the flow of Fairtrade goods between producers and importers. Twenty labeling initiatives and three producer networks are members of FLO, giving a total of 23 members.

The national members of FLO are responsible for licensing the Fairtrade label for use on products that meet FLO's standards, for monitoring the final stages of the supply chain and for promoting the label to

Table 7.3.   Fairtrade labeling initiatives

Fairtrade Austria
Max Havelaar Belgium
TransFair Canada
Max Havelaar Denmark
Association for Promoting Fairtrade in Finland
Max Havelaar France
TransFair Germany
Fairtrade Mark Ireland
Fairtrade TransFair Italy
Fairtrade Label Japan
TransFair Minka Luxembourg
Comerciio Justo Mexico
Max Havelaar Netherlands
Fairtrade Max Havelaar Norway
Asoc. Del Sello de Comercio Justo Spain
Fairtrade Sweden
Max Havelaar Switzerland
Fairtrade Foundation UK
TransFair USA
Fairtrade Labelling Australia/New Zealand

Fairtrade Labelling Organisations (2007).

businesses and consumers in their own countries. Table 7.3 lists the Fairtrade labeling initiatives.

There are many advantages of a common fairtrade labeling system:

- Producers can access fairtrade markets in various countries through a single inspection and certification process.
- The certification system avoids duplication of effort by individual countries and can be operated more cost-effectively.
- All the members of FLO operate to common standards so that products can be more easily sold across national boundaries.

Other Sustainable Foods

Marine Stewardship Council

The Marine Stewardship Council (MSC) (http://www.msc.org) operates an ecolabeling program for sustainable fisheries. The MSC logo is given to fisheries that meet its environmental standard for sustainable and well-managed fisheries. Seafood products from these certified fisheries are given the MSC logo. The logo is most evident in countries where MSC has offices, namely, the United Kingdom, United States, Australia, Japan, and the Netherlands.

The MSC has existed as an independent nonprofit organization since 1999. It was initially founded in 1997 by Unilever, the world's largest buyer of seafood, and the World Wildlife Fund. The MSC environmental standard covers three principles:

1. The condition of the fish stock(s) of the fishery
2. The impact of the fishery on the marine ecosystem
3. The fishery management system

The popularity of the MSC program mainly stems from consumer concern about overfishing and its environmental and social consequences. In March 2008, over 1,400 seafood products in 26 countries had the MSC ecolabel. The MSC standard has been adopted by 26 fisheries, with another 68 undergoing assessment. Over 500 companies have met the MSC chain-of-custody standard for seafood traceability. Almost 10% of the world's edible wild-capture fisheries are now engaged in the program, either as certified fisheries or in full assessment against the MSC standard. This equates to about 4 million tons of seafood.

The Rainforest Alliance

The Rainforest Alliance (http://www.rainforest-alliance.org) was formed in 1987 to conserve biodiversity and ensure sustainable livelihoods by transforming land use practices, business practices, and consumer behavior. With headquarters in New York City, it has offices throughout the United States and worldwide. The organization has three divisions:

(i) The sustainable forestry division
(ii) The sustainable agriculture division
(iii) The tourism division

Food products are covered by the sustainable agriculture division. The Rainforest Alliance standard combines productive agriculture, biodiversity conservation, and human development. Farmers, cooperatives, companies, and landowners who are certified by the Rainforest Alliance must meet social and environmental standards that include agrochemical reduction, ecosystem conservation, and worker health and safety. Farms that meet these standards are given the Rainforest Alliance ecolabel; this is given to farms, and not to companies or products. Certification is provided by the Sustainable Agriculture Network (SAN), a coalition of leading conservation groups.

The Rainforest Alliance agriculture standard was initially developed in 1991 and covers farms that grow tropical crops like coffee, tea, bananas, avocados, oranges, guava, pineapple, vanilla, and cocoa.

Rainforest Alliance standards are mainly adopted by farms in Latin American countries like Brazil, Costa Rica, and Peru. Adoption rates are also rising in countries like Ethiopia, Tanzania, and Indonesia.

In 2007, the amount of Rainforest Alliance–certified farmland surpassed 1 million acres (more than 414,000 ha). Rainforest Alliance standards are adopted by 25,731 farms in 18 countries.

The retail value of Rainforest Alliance–certified products like coffee, bananas, and cocoa reached about US$1.2 billion in 2007. Coffee, one of the major products, has seen sales increase by an average of 93% since 2003.

The success of the Rainforest Alliance ecolabel is largely due to the commitment of large corporations like Chiquita, IKEA, Kraft Foods, Sainsbury's, and Unilever. In the United Kingdom, McDonald's switched its entire coffee supply to certified Rainforest Alliance in 2007. In March 2008, it announced that it would only source Rainforest Alliance–certified tea for its 1,200 fast food restaurants (The Daily Telegraph, 2008). Unilever, the Anglo-Dutch multinational, announced in 2007 that it was converting all its tea for its Lipton and PG Tips brands to Rainforest Alliance–certified sources.

Food Alliance

Food Alliance (http://www.foodalliance.org) is a nonprofit organization that was established in 1997. It initially started as a project of Oregon State University, Washington State University, and the Washington State Department of Agriculture in 1993.

Food Alliance standards include criteria on social and environmental responsibility and product differentiation.

The Food Alliance certification program started in 1998 in Portland, Oregon, with a single apple grower. In 2007, the program had over 275 certified farms and ranches in 19 states in the United States, Canada, and Mexico. These producers manage over 5.1 million acres of farmland, producing beef, lamb, pork, dairy products, mushrooms, dried beans and lentils, wheat, as well as fruits, and vegetables. A growing number of Food Alliance–certified processors are making products like cheese, flour, frozen and canned fruits, and vegetables. The first Food Alliance seafood standards were scheduled to be introduced in 2008.

Sales of certified Food Alliance products were about US$100 million in 2007, with primary products like fruits, apples, and meats comprise most sales.

Carbon Labeling

There has been increasing attention given to the carbon impact or carbon footprint of all activities from heavy industry, to everyday actions, to products specifically. This is due to the growing awareness of climate change and the relationship to carbon impact (being a cause and reduction of carbon impact being a potential solution). Carbon is used as a general means to communicate the impact on climate change—in terms of carbon dioxide equivalents. The methodology to calculate this is a subset of life cycle assessment with a focus on greenhouse gas (GHG) emissions, and all inputs of energy (fossil and otherwise), plus all other sources of emissions. For food, this includes agriculture (arable farming—energy used to manufacture fertilizers; nitrous oxide from fertilizer decomposition; animal husbandry—gastroenteric fermentation in ruminants produces quite large volumes of methane), processing, packaging, and distribution (cooling and freezing—many refrigerants, especially hydrofluorocarbons, contributing to global warming and considerable energy can be used to power refrigerators). Depending on the analysis, use and disposal may or may not be included.

There have been efforts to develop the carbon footprint of food products in order to communicate the climate change impact of a product to consumers, and help them make purchase decisions. However, given that there is uncertainty, or margin of error, in such an analysis, such product labeling has limitations. Further, there has been criticism that such labeling does not reflect meaningful differences a consumer can make.

The leading initiative has been from the Carbon Trust. The Carbon Trust developed a carbon label to provide a measure of a product's carbon footprint (embodied GHG emissions) across its life cycle. The Carbon Trust's carbon label was launched on food and beverage products (orange juice and potatoes) in 2008 in Tesco Stores Ltd in the United Kingdom. The Carbon Trust is in the process of developing a commonly accepted standard for such carbon labels. The aim of the standard is to ensure consistency in measuring carbon footprints (GHG emissions embodied in products and services) and in making and communicating reduction claims. It has three components:

1. A method for the assessment of the life cycle greenhouse gas emissions of goods and services (PAS 2050)—this defines how life cycle GHG emissions of a product should be measured.
2. The Product-Related Emissions Reduction Framework (PERF), which sets out the requirements for making credible claims regarding reduction commitments and achievements on life cycle GHG emissions of products, as measured using PAS 2050.
3. A Product-Related Emissions Communications Guidance (PECG) document to support companies implementing the PAS 2050, or the PAS 2050 and the PERF, to communicate the product-related life cycle greenhouse gas.

Another emerging initiative is carbon calculators. These are tools where consumers provide inputs like personal behaviors or food choices, and their personal carbon impact is then calculated for them. For example, Bon Appétit Management Company, a U.S. food service provider, released a low-carbon diet calculator on the web site http://www.eatlowcarbon.org that provides relative carbon impact information about food choices, based on life cycle research compiled by EcoTrust. This allows a consumer to see how dietary choices can affect their carbon impact. In its facilities it also provides menus based on this information, aiming to educate their customers on how food choice impact climate change and to help in reducing the company's total impact (Bon Appétit, 2007).

Conclusions

The rise in ethical consumerism is leading to an explosion in eco-labels, covering products ranging from foods, cosmetics, household cleaning products, packaging, textiles, furniture, electrical appliances to gardening products. Manufacturers and retailers are adopting eco-labels to address consumer demand for ethical and environmentally preferable products and communicate with these credible and validated tools.

Organic is the dominant sustainable food label. It represents the largest component of the estimated US$65 billion global sustainable food industry. Its widespread adoption is partly because of high investment levels. National governments, intergovernmental organizations, and the private sector have financed organic programs. Unlike other sustainable food label initiatives, its adoption rate spans the four corners of the globe, and in key markets such as North America and Europe, how its use is enforced by law.

Fair trade is gaining popularity because of its impact on third world poverty. Consumers see the fairtrade system as a way of supporting farmers in lesser developed countries. Its major limitation is that demand is mainly from affluent countries, whereas production is in developing countries.

With the interest in carbon labels and other means to communicate sustainable attributes of product, the rise in sustainable labels in the food industry is expected to continue. The success of these programs will help not only differentiate products for consumers to make more responsible decisions, but also offer the opportunity to shift the marketplace to more ethical and environmentally responsible products and practices.

References

BBMG. 2007. *Conscious Consumers Are Changing the Rules of Marketing. Are You Ready?* Highlight from the BBMG Conscious Consumer Report. November 2007.
Bon Appétit Management Company. 2007. *Cooling the Planet.* Available from http://www.bamco.com/PressRoom/press-pre-041707.htm. Accessed September 15, 2008.
Consumers Union. 2002. "Setting the bar for eco-labels." In: *Eco-labels and the Greening of the Food Market Meeting*, November 7–9, 2002. Boston, MA.
DEFRA (Department for Environment, Food, and Rural Affairs). 2007. *An Index of Green Labels.*
Fairtrade Labelling Organisations. 2007. *Annual Report 2006/2007.* Bonn, Germany: Fairtrade Labelling Organisations.
Huber, B., L. Kilcher, and O. Schmid 2007. "Standards and regulations." In: *The World of Organic Agriculture: Statistics and Emerging Trends 2007*, eds H. Willer, and

M. Yussefi, pp. 56–66. Frick, Switzerland: International Federation of Organic Agriculture Movements (IFOAM) and the Research Institute of Organic Agriculture.

Ilbery, B., and D. Maye. 2007. Marketing sustainable food production in Europe: Case study evidence from two Dutch labelling schemes. *Tijdschrift voor Economische en Sociale Geografie* 98(4):507–518.

IRI. 2007. *Times and Trends: A Snapshot of Trends Shaping the CPG Industry: Sustainability 2007.*

Molyneaux, M. 2007. The changing face of organic consumers. *Food Technol.* 61:22–26.

Neuendorff, J. *Certification of Organic Foodstuffs in Developing Countries.* Produced on behalf of the German Society for Technical Cooperation (GTZ) and the Federal Ministrey of Economic Co-operation and Development (BMZ). Available from http://www2.gtz.de/dokumente/bib/02-5121.pdf. Accessed September 25, 2008.

Organic Monitor. 2008a. *The Global Market for Organic Food and Drink.* London, UK: Organic Trade Association.

Organic Monitor. 2008b. *The European Market for Ethical Fruit and Vegetables.* London, UK: Organic Trade Association.

Organic Trade Association. 2007. *Organic Trade Association 2007s Manufacturing Survey.* Conducted by Packaged Facts for the Organic Trade Association. London, UK: Organic Trade Association. Available from http://www.ota.com/pics/documents/2007ExecutiveSummary.pdf. Accessed September 15, 2008.

Scarborough Research. 2007. *When It Comes to Organic Food, the West Is the Best.* Available from http://findarticles.com/p/articles/mi_m5CNB/is_2007_Oct_10/ai_n25013443. Accessed September 15, 2008.

Sloan, E. 2007. New shades of green. *Food Technol.* 61(12):16.

The Daily Telegraph. 2008. *McDonald's Switches to Rainforest Alliance-Certified Tea.* London, UK, March 27, 2008.

USDA (United States Department of Agriculture's Agricultural Marketing Service). *National Organic Program, List of USDA Accredited Certifying Agents.* Available from http://www.ams.usda.gov/nop/CertifyingAgents/CertAgenthome.html. Accessed September 15, 2008.

Willer, H., M. Yussefi-Menzler, and N. Sorensen 2008. *The World of Organic Agriculture: Statistics and Emerging Trends.* Bonn, Germany: IFOAM and Research Institute of Organic Agriculture (FiBL).

Chapter 8

Sustainability in Food and Beverage Manufacturing Companies

Cheri Chastain, J.C. Vis, B. Gail Smith, and Jeff Chahley

Introduction

Food processors represent a concentrated point for controlling the sustainability of the food supply chain. The main direct areas of impact include waste generation, energy use, and water use. Many processors have engaged in improving these direct impacts. For example, Nestle has successfully diverted 95% of its waste from the landfill McCains Foods Limited produces renewable energy to source most of its electricity needs in its plants; and Walkers has reduced its water usage by 50% (FDF, 2008). Implementation of efforts to improve indirect impacts are growing, with focus on improvements in agricultural production, given that it is the major source of environmental impact across the supply chain.

The firs two case studies of food and beverage manufacturing companies provide detail on how leading companies have been successfully implementing sustainability initiatives for direct and indirect impacts, and the fina case study shows how a company begins to incorporate sustainability across its organization.

Sierra Nevada Brewing Company: A Focus on Minimizing Direct Impacts

Sierra Nevada Brewing Company is considered the premier craft brewery in the United States, well known for creating the landmark Sierra

Nevada Pale Ale. Beginning in 1980 in the town of Chico, California, owner Ken Grossman founded this independent brewery that is the second largest craft brewer in the nation and is distributed widely. With comprehensive energy efficien y, recycling, and wastewater programs in place, they are a leader in environmentally responsible business practices.

Energy Efficiency

Sierra Nevada has implemented several means to minimize energy usage. One of Sierra Nevada's key efforts includes alternative sources for energy. Sierra Nevada installed one of the largest fuel cell units (four 300 kW cogeneration fuel cell power units) in the United States to supply electric power and heat to the brewery. Fuel cells use natural or biogas, where hydrogen gas is extracted and combined with oxygen from the air to produce electricity, heat, and water. Waste methane generated at the company's wastewater treatment plant is used as a fuel source for the fuel cell units.

While the power output will produce most of the brewery's electrical demand, the cogeneration boilers harvest the waste heat and produce steam for boiling the beer and other heating needs. Surplus electrical energy is sold back into the power grid.

Their decision to use a fuel cell was based dramatically on lower emissions than conventional power generation, minimal electrical line transmission loss, and the ability to cogenerate and use the waste heat from the fuel cell in the brewing process. In addition to the fuel cells, Sierra Nevada has one of the largest privately owned solar installations in the country with 1.9 MW of solar power going into the plant. The solar panels coupled with the fuel cells produce more than 100% of the plants power needs during peak hours.

Sierra Nevada also installed a vapor condenser to recover waste steam from the kettle-boiling step to preheat process water. The boiler system was modernized to utilize cutting-edge technology with online oxygen sensors and variable-speed blowers to increase energy efficien y and minimize any air emissions. Additional stack-heat recovery equipment has been installed to extract all the useful heat from the system.

Sierra Nevada has also retrofitte fixture with electronic ballast lights and motion sensors, replaced the air compressors with ultraefficient speed-controlled drives, and use high-efficien y motors and refrigeration systems throughout the brewery.

Emissions

The company aims to attain 100% energy self-generation. However, until then, the company participates in a voluntary carbon offset program to reduce emissions associated with electricity purchases, the ClimateSmart Program.

Employees are also encouraged to and rewarded for reducing their emissions by riding their bikes to work, to run errands, or just for fun. This not only helps reduce the consumption of fossil fuels and the level of emissions from cars, but also promotes a healthy lifestyle.

Waste

Reducing consumption and reusing and recycling raw materials is a basic component of the company's operations. Offic paper, cardboard, glass, stretch wrap, plastic strapping, construction materials, pallets, and hop burlap are just some of the many materials recycled at the brewery.

In 2007, Sierra Nevada diverted a total of 1,0 31,038 tons of materials, 98.2% of the total waste from the landfill including 631,681 lb cardboard, 685,874 lb glass, 39,860 lb offic pack/mixed paper, 35,950 lb shrink wrap, and 49,130 lb plastic strapping.

Sierra Nevada has received the WRAP Award (Waste Reduction Awards Program) from the State of California yearly since 2001, and in 2002 was named one of the top 10 recipients of the WRAP of the Year Award for the company's extraordinary waste reduction awareness programs.

Most of the spent brewing materials have a secondary use in the local agricultural community. In conjunction with the Agricultural Department at California State University, Chico, Sierra Nevada provides feed for dairy and beef cows through the spent grain, hops, and yeast it has collected. The surplus spent yeast from fermentation is used as a nutritional supplement for cows, and the compost from the cow manure is used as fertilizer for Sierra Nevada's on-site 8-acre experimental hop field

Another waste recovery initiative is a system installed to recover carbon dioxide that is produced by the natural fermentation process. With this system, Sierra Nevada recovers and recycles most of this gas for use around the brewery and during the bottle-fillin process to assist in dispensing our ales and beers.

Water Conservation

Sierra Nevada made the commitment several years ago to treat all of the production wastewater to remove this burden from the local municipality. A European-designed, two-step anaerobic and aerobic treatment plant was installed to reprocess and purify all the water produced from the brewing operations. Sierra Nevada also focuses on minimizing water use. Currently, they have been able to reduce water usage to almost half of the historical value typically used by breweries in this country. Further, water used for truck washing is collected and purifie at their facility. Sierra Nevada is also installing a system to utilize treated wasts for irrigation on the 8 acre on site hop field

Unilever: A Focus on Minimizing Indirect Impacts

Introduction

Unilever is an Anglo-Dutch company and a global manufacturer of fast-moving consumer goods. About 60% of the €40 billion annual turnover is in food products, about 40% in home and personal care products. Unilever's more than 400 brands include Knorr, Bertolli, Lipton, Good Humor Bryers, Wall's, Becel/Flora, Hellmann's (foods), Dove, Axe, Lux, Sunsilk, Comfort, Domestos, Surf, and Cif (home and personal care). Unilever owns manufacturing operations in almost 100 countries, and runs marketing and sales operations in nearly 150 countries. The company mission is "to add vitality to life": to help people feel good, look good, and get more out of life. Unilever started three sustainability programs around 1995 on fish water, and agriculture. This chapter describes how we in Unilever look at sustainability in the food industry. The views expressed here are entirely those of the authors.

Food Manufacturing and Sustainable Food Supply Chains

Food conservation and preservation techniques ensure we can eat at times of the year or in locations where fresh food is not readily available. And whenever these techniques are used, there are increased opportunities for the food to be traded and made available to people far from the site of production. Food manufacturers work to add value to conserved foods, by creating products that consumers wish to buy and not only satisfy basic safety and nutritional needs but also taste good and feel good, because of their mouthfeel, texture, and cultural context.

For a food manufacturing company such as Unilever, it is important to realize the benefit of modern food conservation and manufacturing technologies, while also addressing as many of the shortcomings of current food supply systems as practical. This means that a wide range of issues—not always mutually compatible—need to be taken into account when making decisions about how *sustainable* any food raw material is. Many of these issues were articulated by the U.K. Sustainable Development Commission in 2002, who define *sustainable food supply chains* as those that:

1. produce safe, healthy products in response to market demands, and ensure that all consumers have access to nutritious food, and to accurate information about food products;
2. support the viability and diversity of rural and urban economies and communities;
3. enable viable livelihoods to be made from sustainable land management, both through the market and through the payments for public benefits
4. respect and operate within the biological limits of natural resources (especially soil, water, and biodiversity);
5. achieve consistently high standards of environmental performance by reducing energy consumption, by minimizing resource inputs, and by using renewable energy wherever possible;
6. ensure a safe and hygienic working environment and high social welfare and training for all employees involved in the food chain;
7. achieve consistently high standards of animal health and welfare; and
8. sustain the resource available for growing food and supplying other public benefit over time, except where alternative land uses are essential to meet other needs of society.

Food manufacturing businesses have direct responsibility for achieving items 1 and 6 of this agenda.

For a food manufacturing business, it is clear that the production of safe, healthy products must be top of the agenda. We decide which foods we produce, and how we market and price these products. We need to develop and promote foods that are safe, nutritious, interesting, and also taste and feel good. We have significant y reduced the fat, salt, and sugar content of many of our foods and will continue to do so in the future—in many cases developing innovative techniques that ensure the resulting products still provide the same level of enjoyment to consumers. We are also developing foods specifi to different parts of the world (e.g., where the diet is deficien in minerals), sectors of the population (e.g., young people who benefi from extra "brainfood"), or to fi in with cultural aspirations (e.g., for a sit-down family meal even when preparation time is short). This is our core business. This has to be our priority.

We also have direct responsibility for ensuring that people who work for us directly have good working conditions and that our factories and other facilities strive to minimize all aspects of pollution.

However, we also make decisions about how, from whom, and under what conditions we buy our raw materials. We therefore have influenc on—if not control of—other actors in food supply chains, and can use this influenc to help create more sustainable supply chains.

Agriculture

Our assessments of our own supply chains, often through life cycle assessments (LCAs)—in common with many academic studies—indicate that the agricultural parts of our supply chain are often areas where there is a great deal of opportunity to improve the sustainability of the entire food chain. The high impacts often result from the use of farm inputs (fertilizer, crop protection agents) that usually have a fairly high energy content and partially end up in the environment (either soil or ground and surface water). This is why Unilever has a sustainable agriculture program.

In agriculture, the three pillars of sustainability can be linked to agricultural yield (in combination with price and quality of course), avoiding negative impacts on environment and instead making improvements—certainly on soil health and fertility, optimizing the use of renewables, and improving the livelihoods of rural communities.

In order to understand and improve the sustainability of the agricultural systems that underlie the production of our raw materials, we based our sustainable agriculture program around four principles (Box 8.1).

Box 8.1 Sustainability principles

Unilever believes that sustainable agriculture should support the following principles:

1. It should produce crops with high yield and nutritional quality to meet existing and future needs, while keeping resource inputs as low as possible.

2. It must ensure that any adverse effects on soil fertility, water and air quality, and biodiversity from agricultural activities are minimized, and positive contributions are made where possible.
3. It should optimize the use of renewable resources while minimizing the use of nonrenewable resources.
4. It should enable local communities to protect and improve their well-being and environment.

Any actions based on these general principles need to be integrated with other business needs such as supply security, food safety, product specifications and cost control. In order to understand how these two sets of requirements can be brought together, Unilever has invested since 1998 in running a number of Lead Agriculture Programmes for a number of key crops around the world.

These programs have all been set up on the basis of the four principles mentioned above. But in order to be able to have a constructive dialog with farmers, the Lead Agriculture Programmes look at 11 different indicators for agricultural sustainability (Box 8.2).

Box 8.2 Sustainable agriculture indicators

1. Soil fertility/health

Soil is fundamental to agricultural systems, and a rich soil ecosystem contributes to crop and livestock performance. Sustainable practices can improve beneficia components of the soil's ecosystem.

Parameters include the following:
1. Number of beneficia organisms (e.g., earth worms per square meter)
2. Number of predatory mites
3. Number of beneficia microorganisms
4. Soil organic carbon (a measure of healthy soil structure)

2. Soil loss

Soil eroded by water and wind can lose both structure and organic matter, diminishing the assets of an agricultural system. Sustainable practices can reduce soil erosion.

Parameters include the following:
1. Soil cover index (proportion of time soil is covered with crop; this protects against leaching and erosion, promotes water binding)
2. Soil erosion (loss of topsoil in percentage per annum or in topsoil/hectare/annum)

3. Nutrients

Crops and livestock need a balance of nutrients. Some of these can be created locally (such as nitrogen), and some must be imported. Nutrients are lost through cropping, erosion, and emissions to the air. Sustainable practices can enhance locally produced nutrients and reduce losses.

Parameters include the following:
1. Amount of inorganic nitrogen/phosphates/potassium applied (per hectare or per tonne of product)
2. Proportion of nitrogen fi ed on site/imported
3. Balance of nitrogen/phosphates/potassium over crop rotations
4. Emissions of nitrogen compounds to air

4. Pest management

When pesticides are applied to crops or livestock, a small but significant proportion can escape to water and air or accumulate in foods, affecting ecosystems and human health. Sustainable practices can substitute natural controls for some pesticides, reducing dependence on synthetic substances.

Parameters include the following:
1. Amount of pesticides (active ingredient) applied (per hectare or per tonne of product)
2. Type applied (profiling positive list, weighting factor)
3. Percentage of crop under integrated pest management

5. Biodiversity

Agriculture has shaped most ecosystems in the world, and biodiversity can be improved or reduced by agricultural practices. Some biodiversity is highly beneficia for agriculture. Sustainable practices can improve biodiversity by *greening the middle* of field as well as *greening the edge.*

Parameters include the following:
1. Level of biodiversity on-site: number of species (such as birds and butterflies) habitat for natural predator systems (such as hedgerows, ponds, and noncropped areas)
2. Level of biodiversity off-site: cross-boundary effects
3. Crop genetic diversity

6. Value chain

Value chain is the term for the sum total of all value-adding activities that lead to putting a product on the market. For food products, farm economics is an integral part of the value chain.

Farmers should develop a f rm grasp of what influence the economics of their farm and what noneconomic value they produce. Sustainable practices should be able to maintain or improve farm economics, and add to nature values and ecosystem service values.

Parameters include the following:
1. Total value of produce per hectare; farm income trends
2. Conformance to quality specifications—nutritiona value, including minerals, pesticide residues, and foreign bodies
3. Ratio of solid waste reused/recycled over solid waste disposed to landfil
4. Marginal costs for various crops and various fields/plot
5. Financial risk management and solvency
6. Nature value and ecosystem service value created

7. Energy

Although the energy of sunlight is a fundamental input to agriculture, the energy balance of agricultural systems depends on the additional energy supplied from nonrenewable sources to power machinery. Sustainable practices can improve the energy balance and ensure that it remains positive—there is more energy coming out than going in.

Parameters include the following:
1. Balance: total energy input/total energy output, including transport where relevant
2. Ratio renewable over nonrenewable energy inputs
3. Emissions to air (greenhouse and pollutant gases)

8. Water

Some agricultural systems make use of water for irrigation; some pollute or contaminate ground or surface water with pesticides, nutrients, or soil. Sustainable practices can make targeted use of inputs and reduce losses.

Parameters include the following:
1. Amount of water used per hectare or tonne of product for irrigation
2. Leaching and runoff of pesticides to surface and groundwater
3. Leaching and runoff of nitrogen/phosphates/potassium (nutrients) to surface and groundwater

9. Social/human capital

The challenge of using natural resources sustainably is fundamentally a social one. It requires collective action, the sharing of new knowledge, and continuous innovation. Sustainable agriculture practices can improve both social and human capital in order to ensure normal outputs. The prime responsibility for this should remain with the local community, leading to realistic and actionable targets.

Parameters include the following:
1. Group dynamics/organizational density (farmer groups)
2. (Rural) community awareness of relevance and benefit of sustainable practices/connectivity to society at large
3. Rate of innovation

10. Local economy

Agricultural inputs (goods, labor, services) can be sourced from many places, but when they come from the local economy, the expenditure helps to sustain local businesses and livelihoods. Sustainable agriculture practices can help to make the best use of local and available resources in order to increase efficien y.

Parameters include the following:
1. Amount of money/profi reinvested locally
2. Percentage of goods/labor/services sourced locally
3. Employment level in local community

11. Animal welfare

Animal husbandry systems are becoming ever more specialized and therefore further removed from the wild habitat where farm animal ancestors evolved. Treatment of animals in contemporary artificia environments is a major ethical concern. Care must be taken so that the animals can live in harmony with their environment.

Parameters include the following:
1. Feeding, housing, watering
2. Treatment of diseases
3. Freedom from abuse

All Unilever's Lead Agriculture Programmes involve groups of farmers or plantations (for tea and palm oil). Together with advisors, either from Unilever or from academia and NGOs, they work on identifying good agricultural practices in order to bring the sustainability values of each indicator to a higher level. Rule number one is that the sustainable practices must have a demonstrable benefi for the farmer (and the farm). Although an old saying is that "farmers live like they could die tomorrow, but farm like they will live forever," in practice many farms are not managed in a sustainable way. For conventional farmers in developed countries, the main shortcomings are a heavy reliance on agrichemicals, a disregard for soil fertility, not enough attention for biodiversity, and little understanding of the social dynamics of rural communities. Many farmers use more agrochemicals than necessary and even more than recommended. As one farmer put it: "I do not want my harvest to be limited by nitrogen shortage, so I usually apply twice as much (fertilizer) as recommended." In order to introduce integrated pest management techniques, farmers usually must learn how to recognize a significan outbreak of the key pest insects and diseases for each of their crops (in each microclimate) and what to do when it happens, rather than relying on reducing the risks of outbreaks by prophylactic spraying with pesticides. Sustainable farming is knowledge-intensive, not chemical-intensive. Many lessons can be learned from organic agriculture. But to do away with synthetic fertilizer and pesticides altogether (which is a requirement in organic farming), changes the risk profil of farming dramatically and usually has a negative effect on yields. Which is why Unilever drives toward integrated pest and crop management, but believes that for mainstream agriculture, moving to organic farming is too risky and not really necessary.

Different types of supply chains call for different types of intervention. Where Unilever buys directly from farmers (e.g., for tea, tomatoes, gherkins), Lead Agriculture Programmes can be set up through Unilever supply management groups. But where we buy through intermediate traders and processors, we have to rely on cooperation with these suppliers. A challenge in themselves is the supply chains for the global commodities, such as soybeans, palm oil, and sugar. Unilever participates in a number of commodity-specifi sustainability initiatives, such as the Roundtable for Sustainable Palm Oil and the Roundtable for Responsible Soy.

In an attempt to combine farmer's interests with food manufacturers requirements, we have set up our Lead Agriculture Programmes on a crop-by-crop basis. For farmers, a farm-based approach would, of course, be much more interesting. Annual crops are grown in rotation with other crops, after all, and there are many ways in which crops in a rotation affect each other. Leftover stubble in the fiel can be plowed in and provide organic matter and nutrients for the next crop. Deep-rooting crops can leave a suitable seed bed for the next crop. But crop residue in the fiel can also lead to massive nitrate leaching, for example, in the case of leguminous crops (which fi nitrogen from the air) such as beans and peas. In such cases, sowing in an early winter crop as a cover crop can be a way to store nitrogen until the next spring crop can be sown. In order to bring as much crops as possible within reach, Unilever therefore participates in a food industry initiative on sustainable agriculture, the SAI Platform (Sustainable Agriculture Initiative Platform).

Table 8.1. Typical features of four types of food supply chains

	Type of food supply chain			
	Local	Conserved	Manufactured	Commodity
Overall complexity of supply chain	+	++	+++	+/++
Transportation distance ("food miles")	+	++	++/+++	+++
Number of processing steps	+	++	+++	+
Storability of finishe product	+/++	+++	++/+++	+++
Size of market for finishe product	+	++	++	+++
Seasonality of finishe product on market	+++	+	+	++
Volatility of market price	+/++	+	+	+++
Demand for further processing by end user	++	+/++	+	++/+++

+, low; ++, medium; +++, high.

SAI Platform aims to be a competency center for the food industry in the area of sustainable agriculture.

The insights gained from our Lead Agriculture Programmes, and partnerships and dialog developed as part of joint initiatives in the area of sustainable agriculture have enabled Unilever to make a start on an ambitious program to help improve the sustainability of the agricultural systems that supply our raw materials.

But it is clear that commercial interests, however important (see Table 8.1), will never be sufficien to drive the radical overhaul in food supply systems that the *sustainability* and *sustainable development* agendas require. Many different actors need to address many different challenges (see Table 8.2). Among the biggest challenges, in our view, are the position of small farmers and the overall yield gap in agriculture.

Every agronomist knows that there is a difference between theoretical yield and actual yield. Research plots deliver better results than implementation trials, and implementation trials deliver better results than when the crop variety reaches the average farmer. There are many reasons for this, ranging from trial locations on good soils with fla land and good climatic conditions, choosing optimum varieties for

Table 8.2. Typical responsibilities assigned to actors within food supply chains

U.K. Sustainable Development Commission priorities	Actors within supply chain					Outside chain	
	Farmers and growers	Transport and distribution[a]	Processing and manufacturing	Retailing	Consumers and citizens	Governments	Research and development
1. Safe, healthy products, nutrition, and information	++	++	+++	+++	+	++	++
2. Rural and urban economies and communities	+		+	+		+++	
3. Viable livelihoods from sustainable land management	+		++	++		+++	+
4. Operate within biological limits of natural resources	++				+	+++	+++
5. Reduce energy consumption, minimize inputs, renewable energy	++	+++	+++	++	+	+++	+
6. Worker welfare, training, safety, and hygiene	+	+	+++	+++		+++	+
7. High standards of animal health and welfare	+++	++	++	++		+++	+
8. Sustaining the resource	+					+++	++

[a] Includes transport and distribution both before and after primary processing and manufacturing.

+, low; ++, medium; +++, high.

microconditions, to spending time on monitoring the crop in the fiel , to simply paying attention. After all, there are good farmers and bad farmers. But, with a growing world population and a growing realization that one day the world will have to rely on agriculture for food, feed, fibe , and fuel, there is a clear need to make the yield gap smaller than it is today. For sake of brevity, here is one example of such a yield gap (but there exist many). For palm oil, the average yield per hectare per year in the main producing countries, Indonesia and Malaysia, is in the order of 3 tons (and has been 3 tons for the last 10 years). But in the best managed commercial plots of well-managed plantation companies, yields reach 6–7 tons/ha/year, and in some research plots yields can reach 9–10 tons/ha/year. This yield gap, in combination with the disadvantaged position of small farmers described above, represents the untapped potential of agriculture in the world. It explains why scenario studies of agricultural yields, even for Europe, reveal time and again that the total harvest realized in Europe could also be realized on only 20% of the land, using only 20% of the fertilizer and 10% of the pesticides, if farming could take place on land best suited to the crop in question, in the right agroecological zone, with the best varieties, the best fertilization, best irrigation, and the best crop protection programs. Although the numbers are simplifie here, the overall outcome is true. It may be hard to believe this, for non-agronomists. A simple example from our own Unilever programs may be convincing. It has proved possible to increase yield of processing tomatoes in Australia from 40 tons/ha to over 100 tons/ha, while reducing the need for irrigation water by 40% and the need for pesticides by 85%, by switching from furrow irrigation to drip irrigation. The example is from several years ago, and people who know about this might say, "Of course, drip irrigation for tomatoes in Australia is much better than furrow irrigation." We know that now. But we did not know that 15 years ago, and the farmers did not know this 15 years ago, and that is the point that matters.

Unleashing the untapped potential of small farmers and narrowing the yield gap, without further damage to environmental or social capital, are probably the biggest challenges of sustainable agriculture programs in the world. Finding ways to use standards as tools for development, rather having them turn into nontariff barriers to trade, might provide the solution.

Kraft Foods: A Focus on How to Begin to Implement Sustainability across the Organization

Introduction

With more than US$34 billion in annual sales, Kraft Foods is one of the largest food and beverage manufacturers in the world. For over 100 years, Kraft has been offering foods that fi the way consumers live. Kraft markets a broad portfolio of iconic brands in 155 countries, including seven brands, with revenue exceeding US$1 billion, including *Kraft* cheeses, dinners, and dressings; *Nabisco* cookies and crackers; *Oscar Mayer* meats; *Philadelphia* cream cheese; *Post* cereals; *Jacobs* coffees; and *Milka* chocolates. Kraft products are present in over 99% of all U.S. households, and millions of times a day in over 155 countries, consumers are reaching for their favorite Kraft product.

Kraft recognizes that its decisions have a sizable impact on the earth's resources. With the heightened focus on global warming, water scarcity, biodiversity, social issues in developing countries, etc., Kraft has recently taken sustainability to a key business strategy. At Kraft, the vision for sustainability is that *"sustainability is a part of every decision we make."*

Like many other organizations, Kraft approaches sustainability by focusing on economic, social, and environmental responsibility elements, as outlined in Figure 8.1.

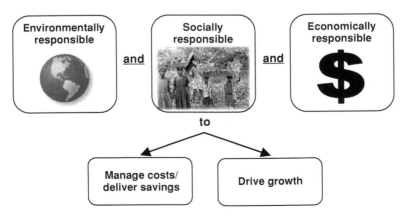

Figure 8.1. Undertake initiatives that are environmentally, socially, and economically responsible.

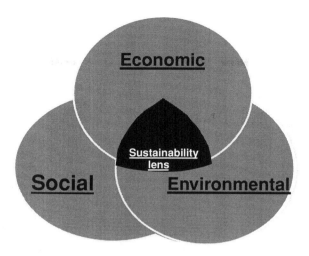

Figure 8.2.   Sustainability lens.

In order to have a sustainable solution, one must have all three el-ements of economic, social, and environmental responsibility present and all must be impacted in a neutral or positive manner. If one element is impacted negatively, then it is not a sustainable solution. To illus-trate this point further, Kraft has adopted the concept of a sustainability lens. Figure 8.2 illustrates the intersection of all three elements as being the desired solution point when making sustainability a part of every decision made.

When assessing initiatives through the sustainability lens, one is con-sidering long-term impact of actions by looking near in as well as into the future and employing *end-to-end* thinking (life cycle analysis). This model assumes there is a belief that achieving long-term growth de-pends on innovation being reignited by looking at the world and your business model through a different set of filters

With the sustainability lens in mind and going one step further into more specifi sustainability focus areas, Kraft has developed what is called the *sustainability wheel*. Figure 8.3 illustrates the concept of looking at various aspects of the supply chain from both a design as well as a continuous improvement perspective. One can think of the center of the wheel as the starting point for a new initiative, designing in sustainability wherever possible, and then focusing on continuously improving the original design.

Figure 8.3. Kraft sustainability wheel.

When beginning a new project one builds sustainability into the design up-front and then drives continuous improvement throughout the supply chain from "end to end" thereafter.

Let us look more closely at the Supply Chain elements of the wheel, which includes energy, water, waste, and transportation/distribution.

In 2006, Kraft consumed more than 2.8 billion mega joule of energy, 56 million cubic meter of water, and generated over 1 million tonnes of solid waste and 2.8 million tonnes of greenhouse gases in the manufacturing of just over 9 billion pounds of food products. In order to think about these inputs and outputs in a simple manner, Figure 8.4 shows the Supply Chain supply chain sustainability model.

On the left side of the GSC sustainability model are the basic inputs to making food: raw and pack materials, water, fossil fuel–based energy,

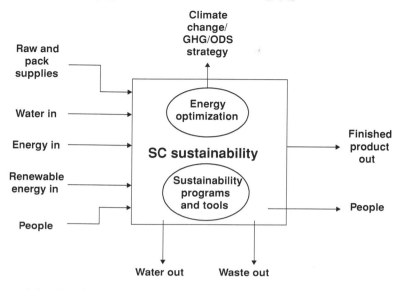

Figure 8.4.   Supply chain sustainability model.

renewable energy, and of course people. On the top of the model are the emission wastes that contribute to greenhouse gases and ozone depletion. On the bottom of the model are the wastewater and solid waste streams, which may include residual wastes in the form of energy. For example, if a product is fille at a high temperature and then cools in a warehouse, the product releases residual wasted energy. Waste-out also includes waste that can be recycled or reused (Kraft recycles over 88% of its waste stream from its manufacturing facilities). On the right side of the model is the desired finishe product that is saleable to customers and consumers as well as the people who helped to produce the products and then return to their homes each day. Within the supply chain Sustainability box are those things that occur to minimize the wastes and inputs and maximize the desired finishe product out. Through the insights of David H. Gustashaw of InterfaceRaise, a division of InterfaceRaise Inc. and pioneers in sustainable development, Kraft has learned to think about sustainability in these terms and has also learned the concept of *mass efficiency*. It is merely your ability to convert your inputs into desired outputs, while minimizing waste. This concept was deployed across Kraft manufacturing facilities with the intent to baseline how

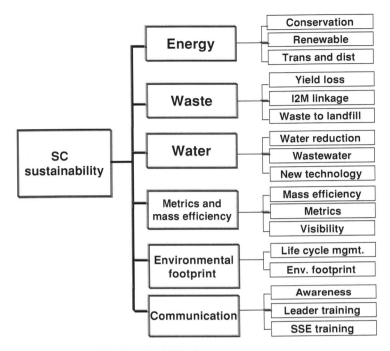

Figure 8.5. Supply chain sustainability focus areas.

efficien plants are in making food today and to establish improvement goals for the future.

With a supply chain sustainability model, Kraft has leveraged it to structure how it will drive continuous improvement in sustainability across the supply chain. Figure 8.5 outlines the work structure and work teams that Kraft has adopted.

Supply Chain Sustainability

In order to tackle the six key supply chain focus areas of sustainability, a champion has been assigned to each area and empowered to lead a cross-functional, global team whose mission is to defin the strategies, tools, and where appropriate providers of tools and processes that will take Kraft to higher levels of performance in sustainability. Some of the focus the teams have are on near-in needs that are required to create or strengthen the sustainability foundation, such as sustainability metrics processes, communication and education plans to engage over 90,000

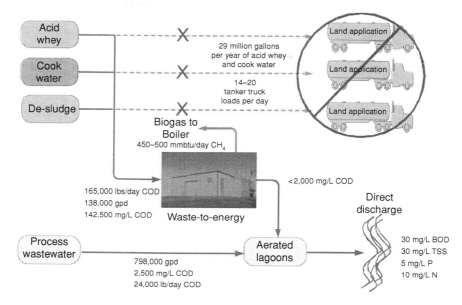

Figure 8.6. Anaerobic digestion project at one facility.

employees, and tools for assessing impacts on the environment from cradle to cradle or farm to fork. Others such as energy, water, and waste will have a longer tenure and will continue to search for new continuous improvement technologies, practices, and solutions.

Energy
Kraft is targeting to reduce energy consumption by 25% by 2011. This is no easy task, though with the help of cross-functional, global energy teams working on energy conservation, transportation optimization, and exploitation of alternative and renewable energy sources, it appears to be achievable.

An example of a sustainable solution recently put into place is the use of anaerobic treatment systems in cheese-making plants, as illustrated in Figure 8.6. Two Kraft plants located in New York State are benefitin from such technology. In the past, these plants produced a liquid waste stream that consisted of acid whey, a by-product of cheese making. Traditionally, this acid whey has been captured, loaded into tanker trucks, and then hauled to local farms where the waste was land applied on local farms. It was truly an unsustainable solution. The anaerabic treatment system is based on mobilized fil technology from a company called Ecovation, where the acid whey stream is separated from the liquid waste stream, fed to a bacteria-enriched media, and then a biogas is created that results from the bacteria digesting the whey by-product. This gas is then sent to the boilers within the facility and used to create energy for the plant, offsetting roughly 30% of the plant's requirements. In addition, the filtratio of the effluen results in a cleaner wastewater stream that reduces the demands on wastewater treatment. The new technology has created jobs in the community for the construction project, reduced dependence on fossil fuels, eliminated the need to land spread whey helped in the improvement of wastewater quality, and it is an economic improvement.

Kraft's global energy team is also evaluating alternative and renewable energy strategies, including waste to energy and nature to energy. Waste to energy includes solid waste, food waste (e.g., meats), and liquid waste such as the anaerobic treatment project mentioned earlier. Nature to energy includes solar energy, wind power, hydropower, and geothermal considerations.

In the area of transportation optimization, there are a number of activities going on, including active participation in the SmartWay Transport Partnership, a U.S. EPA-sponsored program that "is an innovative collaboration between the U.S. EPA and the freight industry to increase energy efficien y while significant y reducing greenhouse gases and air pollution" (EPA).

Some of the concepts being adopted or explored by the Kraft-owned transportation vehicles include the following:

1. Use of nitrogen for tire inflatio
2. Use of low-rolling resistance tires
3. Exclusive use of 53 ft trailers
4. Implement use of *smart reefer* technology
5. Full implementation of auxiliary power units in all tractor specifi cations
6. Use of truck-stop electrification—Idleair
7. Development of standard equipment specificatio using aerodynamically designed bodies
8. Use of single-wide tire technology
9. Implement use of fuel additive—cetane booster
10. Testing of advanced engine oil filtratio system

All these initiatives are helping to increase fuel efficien y and reduce greenhouse gas emissions. In addition to truck fuel-efficien y optimization, there is a separate initiative that is focused on minimizing food miles by leveraging intermodal transportation, reducing empty back-hauling by rerouting trailers to other pickup points, leveraging cooperative routing arrangements with other shippers, and maximizing cube and weight on loads. Together, these initiatives will significant y lower emissions resulting from the transportation of food.

Waste

The second primary waste stream leaving food manufacturing facilities is solid waste. A lot of improvement can be done with an increased focus on waste recycling in the facilities through internal recycle collectors, recycle compactors, etc. Other avenues for reducing waste to landfil are to try and fin higher purposes for the waste. This includes using edible food waste as a by-product ingredient for other things such as other Kraft products, animal or pet food, as well as a source of fuel to create waste to energy. A good example of this is the burning of coffee chaff, a waste component associated with roasting and grinding coffee beans, in boilers within the manufacturing facilities. This activity eliminates the need to haul the chaff away from the facility and dispose it, while providing an alternative fuel that displaces the need for fossil fuel consumption.

A second example of how solid waste is being dealt with more responsibly is the partnering with key suppliers on the capturing and reusing of packaging materials. One such vendor is Sonoco, a key supplier for a wide range of packaging supplies, who is also very committed to sustainability. Sonoco has been working with our manufacturing facilities, and the employees within to establish packaging material reclaim programs that allow Sonoco to capture and remove almost all of the packaging material waste generated. Sonoco then either reintroduces the packaging waste into their manufacturing process or incinerates it and generates energy from waste, thus reducing the need for fossil fuels and helping Kraft to reduce its burden on landfills It is a great example of how partnering with others within the entire cradle-to-cradle life cycle can benefi everyone.

The other obvious area for focus is to not generate waste in the firs place. Within the box in the supply chain sustainability model in Figure 8.4 is a focus area called *sustainability programs and tools*. One of those programs is yield improvement and focusing on all of the losses that occur in the manufacturing of food products. Common tools deployed by continuous improvement teams include mass balances, process mapping, other lean manufacturing/six sigma tools, and reliability improvement tools. This approach to waste elimination has now been expanded to drive out waste across the supply chain, end to end, looking at reasons for sales returns, product not suitable for sale, and damage. Another process that is currently being rewired is the idea-to-market (I2M) process, which is an elaborate set of procedures, gates, milestones, reviews, etc., required to successfully launch new products or product/packaging changes. Within it there now exists specifi questions related to sustainability, design for sustainability, packaging material considerations, and a host of other things to consider when innovating new ideas. Thus, by designing out waste one can reduce waste.

Water

Many would argue that fresh water scarcity is quickly becoming the world's next global warming problem. Although the cost of fresh water is quite inexpensive in some parts of the world, such as North America, it is an expensive material within the manufacturing process. The cost to pump it, heat it, cool it, and treat it makes wastewater seem more like "gold water." The water leaving a facility has a lot more intrinsic value than the water that entered the facility. In addition to the cost of water, the fact that many facilities are located in rural areas means that facilities need to be mindful of the water availability in local aquifers and do things to ensure the aquifers remain viable.

The current global water sustainability team is identifying instances where facilities may use water as a cooling medium and simply bring water in one side of the plant, pass it through heat exchangers as a cooling media, and then discharge it back into the wastewater stream. By carefully looking at all the potential uses of water using water mass balances, and understanding the points of heating and cooling, water is now being used in a more elaborative and well-thought-out manner. Once water is finishe doing its part within the facility, a fina challenge is done to see what can be extracted from it and reused in the facility instead of going to wastewater.

For example, in order to retain high-quality standards and to ensure superior food safety, the food manufacturing equipment must be sanitized on a frequent basis, using caustic and alkali chemicals mixed with water, something called clean in place or CIP. In the past the CIP-rinse water was neutralized using additional chemicals prior to it being sent out as wastewater. Costly chemicals were simply going down the drain along with the food elements removed from cleaning of the equipment. Now, with the help of advancements in technology, most of the chemicals can be removed from the wastewater and safely reused for further cleaning. Additionally, work was done with chemical suppliers and Kraft quality resources to optimize the sanitation cycles and thus minimize overuse of chemicals in the firs place. Overall, these efforts result in reduced water usage, reduced chemical usage, and the transportation emissions associated with getting them to facilities, and reduced loading on the wastewater treatment facilities.

Metrics

A cross-functional global sustainability metrics team was assembled that leverages external experts and key reference materials such as *GHG Protocol* by World Resource Institute and World Business Council for Sustainable Development. Some of the key challenges include trying to figur out what data to track, which components of the supply chain to include, how frequent to collect the data, and doing so in a way that is easy to do, simple to understand, and possible to do in every country and language around the world. This includes getting consensus on definition as well as units of measure. In the end, the team aspires to have data that is visible internally to everyone for benchmarking purposes, have processes in place that are necessary to validate metrics and align with usages invoiced from suppliers, eliminate the numerous one-off spreadsheets that feed the corporate data requirements today, and be able to provide data information in a reasonably painless manner, including those required for any carbon-based reporting regulations or other external communication programs.

The global environmental footprint team piloted a Life Cycle Analysis (LCA) on one key product and learned that it takes a lot of time and effort to get data at a stock-keeping unit level, especially when the manufacturing facility produces various products and formats, yet only has 1 meter to measure energy in or water into a facility. The team also evaluated the pros and cons of the various LCA software tools available and talked with industry experts on their experiences with the tools. Team members have been attending sustainability and environmental footprinting-related conferences to learn more and to network with others who have already been engaged in LCA activities. The language, terminology, and tools are clearer now, and providers have been assessed with a select few identifie to assist on an as-needed basis, and processes are developed to engage the business owners in sustainability. The team has also learned that LCAs are one aspect of a larger topic called life cycle management and life cycle thinking.

Communication

When launching a program, it is always important to have a solid communication and education strategy in place along with appropriate reinforcement and recognition of desired behaviors and results. An organization must be informed about the following:

1. What exactly is sustainability and what is the company's vision for the initiative?
2. What is the action plan for my function (or our group) to help achieve the vision?
3. Who are the key resources working on it and whom do I go to if I have questions or ideas?
4. What is my role and what skills will I need in order to complete the work?
5. When will I be trained and what is in it for me?

If you happen to miss one or more of the above items, then your chances of managing change successfully are significant y diminished. For that reason, a communication team was commissioned to develop a game plan that addresses the above points and allows sustainability to be rolled out successfully to Kraft worldwide. The team's focus areas included leadership training for salaried employees, sustainability training modules to allow for adaptation to existing hourly employee safety, security and environmental training programs, and final y overall communication and awareness planning that leverages the various communication vehicles that exist within the company.

Key communication activities have included sustainability overviews to engage Kraft executives, marketing and sales teams, supply chain teams, research and development teams, information system teams, etc., in the sustainability agenda. These reviews focused on definin sustainability, outlining Kraft's sustainability vision, strategies, action plans, and a review of key sustainability initiatives and examples that are occurring across the company.

References

EPA (United States Environmental Protection Agency). *Smartway Transport*. Available from http://www.epa.gov/smartway/transport/index.htm. Accessed September 15, 2008.

FDF (Food and Drink Federation). 2008. *Environmental Company Case Studies*. Available from https://www.fdf.org.uk/environment_casestudies.aspx. Accessed September 15, 2008.

Chapter 9

Sustainability in Food Retailing

Cheryl Baldwin

Introduction

Food retailing spans a range of operations from grocery stores, super-stores, warehouse stores, to natural food stores, with sales predominat-ing from grocery stores and superstore formats (Davies and Konisky, 2000). Food retailers have been exposed to sustainability through their suppliers' initiatives for years. Now that more suppliers are engaged in the issue, more retailers are taking proactive approaches themselves. This is a positive direction since food retailers are in a position in which they can influenc behaviors of suppliers and consumers (Davies and Konisky, 2000).

At this point, most major retailers have initiated sustainability ac-tivities from carrying store-brand organic products to using renewable energy. A 2007 survey of consumers in the United States perceived that the top three green brands across all consumer brands were all from food retailers—Whole Foods, Wild Oats, and Trader Joes (Lan-dor, 2007). Further, consumers currently identify Whole Foods Market and Wal-Mart as socially responsible companies (BBMG, 2007).

The trade associations are also taking action. The Grocery Manufac-turers Association (GMA) held its firs Environmental Sustainability Summit in January 2008. At that meeting, GMA directed its sustain-ability focus on energy, water, and packaging within companies. The Food Marketing Institute began a Sustainability Task Force in 2006 and identifie how the association may facilitate retailer's and wholesaler's integration of sustainable practices. Their primary outcomes were tools to help a retailer identify its baseline sustainability strategy.

Life Cycle Impacts of Food Retailing

The life cycle impacts of food retailing has had limited study. The direct impacts of operations are energy use, waste production, distribution, and influenc on consumer choice (Davies and Konisky, 2000). However, the majority of the total impact of a retail operation goes beyond the direct impacts (within the store's control) and comes from their supply chain.

The United States Department of Energy estimates that food retailers are the largest commercial user of refrigeration (DOE, 2006). Energy use is predominated by refrigeration requirements, which comprise over half of energy use at retail outlets (Davies and Konisky, 2000). The other sources of energy use are lighting and heating/air-conditioning of the building. The result is that food retailers comprise 4% of all commercial building energy use in the United States (DOE, 2006). As a result, efficien use of energy should be a priority of food retailers.

The Stop & Shop Supermarket Company, LLC, which includes Stop & Shop stores in New England, New York, and New Jersey, and Giant Food stores in the Mid-Atlantic, has gained the United States Environmental Protection Agency (EPA) Energy Star Leader award, the only supermarket chain in the country earning the recognition (Stop and Shop, 2008). Energy efficien y measures include skylights and natural lighting in stores, high-efficien y lighting, refrigeration systems with variable—speed compressors and low-energy glass doors, occupancy sensors, and reflect ve, insulation roofin (Stop and Shop, 2008). The company has also committed to increasing the energy and fuel efficien y of its flee of 165 trucks (Stop and Shop, 2008).

Complete conversion to chlorofluorocarbon (CFC)-free refrigeration will reduce additional environmental impacts with the high use of refrigeration. CFCs have been used in refrigeration, but are known ozone-depleting substances. The EPA regulations issued under Sections 601-607 of the Clean Air Act have phased out the production and import of ozone-depleting substances (ODS), consistent with the schedule developed under the Montreal Protocol. The phaseout has operated by reducing, in stages, the amount of ODS that may be legally produced or imported into the United States. Now that zero new production/import is allowed, ODS users have been converting to alternatives. For example, GreenWise, a store in Florida owned by Publix, has reduced its use of ODS by using more environmentally preferable, secondary coolant systems and refrigerants (McTaggart, 2008). Additionally, the EPA established a recognition program in 2007, called GreenChill Advanced Refrigeration Partnership, for retailers that are adopting technologies, strategies, and practices that reduce emissions of ODS and greenhouse gases and increase refrigeration system energy efficien y.

Waste from food retailers is composed of packaging and food. Packaging comprises the majority of waste from most food retailers. Packaging is in the form of cardboard, paper, plastic, steel, aluminum, glass, and wood (Davies and Konisky, 2000). All of these components can be recycled, diverting the waste from the landfill Retailers are taking efforts to recycle more packaging material, for example, in 2007, the United States retailer Wegman's recycled 1.5 million pounds of plastic bags (returned by customers) and 1.5 million pounds of plastic stretch/shrink wrap (Wegmans, 2008). The uses of this recycled plastic include production of new bags and durable composite materials for deck and railing (Wegmans, 2008).

In addition to recycling, efforts to reduce total packaging can further aid in efforts for waste (and resource) reduction. For example, bulk items can reduce packaging. A study of a Michigan retailer found that the sale of bulk items significant y reduced solid waste (Keoleian et al., 1991). The advantages of bulk items further include lower costs of inventory, display space, and labor for handling and stocking compared to prepacked items, even though bulk products must be weighed and priced at checkout (Keoleian et al., 1991).

Often secondary packaging is a source of extraneous material use. In a study by Franklin Associates (1995), it was found that the total amount of secondary packaging waste has increased, and can be reduced. Reusable containers can be used, such as injection molded plastic crates for carbonated beverages and flui milk and plastic cases for fresh produce. Plastic pallets provide another reusable delivery option. These products can be used for 12 years and are lighter than wood to reduce fuel use during distribution (Moore, 2008). Further, at the end of life, the plastic pallets can be recycled into new pallets.

The fraction of waste from food is dependent on the type of retailing operation; however, it is typically a lower component to the total than packaging. Food waste is often due to retail pressures of high standards for size, shape, color, and uniformity. Final food waste can be composted or donated to reduce total landfil contribution and increase utilization of the resources imbedded in the food waste. Publix has begun to compost food waste in Florida, in a partnership with the Florida Atlantic University (McTaggart, 2008).

Distribution can be considered from the point of view of getting the food to the store and from the point of view of getting consumers to the store. In many circumstances, the total life cycle impact of a food product may be outweighed by the impact of the consumer driving their car to the store. This has been seen when consumers travel to the store for the sole purpose of purchasing a few items, and not conducting other errands on the same trip (Braschkat et al., 2003). Careful site location may help in reducing this impact, choosing sites close to and easily accessible to population centers. Consumer education and programs on maximizing trip efficien y may help. Other store-based programs like delivery can reduce fuel use during the distribution of the product from the store to the consumer (and also can be done with more fuel-efficien or alternative-fueled vehicles).

Distribution of food to the store has already been discussed in Chapter 3. Food retailers are increasingly involved in wholesaling and distribution, rather than relying on wholesalers or distributors. This level of vertical integration can maximize transportation efficiencies The retailer can also control the type of fuel used to move toward the use of alternative fuels, flee maintenance to ensure fuel efficien y is maximized, and routes taken to keep distance at a minimum. This type of vertical integration has also moved into food processing. For example, Safeway produced half of its private label products in 1998 (Safeway, 1999). This has potential for greater efficiencies

Food retailers can influenc consumer behavior with education, programs, and item stocking. In a study by GfK Roper Public Affairs and Media, retail grocers ranked highly as a source of information about food choices for consumers (Sustainable Food News, 2007). With effective execution, the retailer's upstream effect can reach to the agricultural impacts of the food supply. As covered in other chapters in this book, agriculture is often the predominate source of life cycle impacts. With produce being a key competitive differentiator and growth driver for retailers, as customers increasingly look for fresh and healthy foods, this area has high potential for a positive impact when education is done effectively (SUPERVALU, http://www.supervalu.com). Further, in a survey conducted with stakeholders in the United Kingdom, it was found that nonretailers feel that more communication about sustainable choices in the store needs to be done (Opinion Leaders, 2007). Point of purchase education, including signs and certificatio logos, has been found to be effective means to provide consumers with information (Davies and Konisky, 2000).

For example, GreenWise stores, owned by Publix, provide unique shelf tags to help consumers identify between products that are all natural, 100% organic, and those made with organic ingredients (McTaggart, 2008). This is an area of tremendous opportunity for retail differentiation and impact.

What Retailers Are Doing

Wal-Mart Stores, Inc.

Wal-Mart Stores, Inc. has 7,323 Wal-Mart stores and Sam's Club locations worldwide. They have been widely publicized for their initiatives in sustainability because of their size and scope, serving 176 million consumers weekly (http://walmartstores.com/). They have stated goals of creating zero waste, supplied by 100% renewable energy, and provide products to their consumers that sustain resources and the environment. While, Matt Kistler, Senior Vice President for Sustainability, said, "Today we do no have clear-cut direction of how we're going to attain every goal" (Greenbiz.com, 2008), they have made progress during the short time they have been engaged in these issues.

The company has stated that its direct impacts span only about 8% of its footprint, with the remaining 92% coming from its supply chain (Greenbiz.com, 2008). To green its supply chain the company launched a packaging scorecard last year, covered in Chapter 4. This scorecard is part of Wal-Mart's aim to achieve its goal to reduce packaging by 5% by 2013 and creating zero waste. Although suppliers were supposed to provide packaging information on all products by the end of February 2008, Wal-Mart only received information for about half of its products in April 2008 (Greenbiz.com, 2008). To further the achievement of the zero waste goal, they have established a mid-term goal of reducing waste by 25% by October 2008 (http://walmartstores.com/). Packaging reductions will help, but the company is relying on recycling any remaining solid waste.

To reach their renewable energy goal, Wal-Mart has been working to make its existing stores 20% more efficien in 7 years, building new stores that are 30% more efficien in 4 years, and getting the truck flee 25% more efficien in 3 years, double in 10 years. Further, they are working on many solar projects, and in 2008 they are planning to start using hybrid trucks. Wal-Mart has already made its flee 20% more efficient compared to 2005, by designing aerodynamic trucks and using auxiliary power units that turn off the engine, but not the heating, cooling, and lights (Greenbiz.com, 2008). In the stores, they have found efficiencie by incorporating secondary loop refrigeration systems, using natural light, light-emitting diode (LED) lighting, and reflect ve roofs.

To provide sustainable products to its consumers, under its private label Sam's Choice brand, the company began offering three Fairtrade-certifie coffees, one Rainforest Alliance–certifie blend, and one USDA organic coffee (Greenbiz.com, 2008). The coffees are roasted by Cafe Bom Dia, a Brazil-based company that offsets its emissions through CarbonNeutral. These product offerings were part of the 2008 Earth Day promotion of environmentally preferable products. For the 2008 Earth Day promotion, Wal-Mart featured more than 50 products in stores and 500 online, from transitional cotton shirts to mulch made from rubber to environmentally preferable cleaning products.

SUPERVALU

SUPERVALU is a retail network in the United States, headquartered in Minneapolis, Minnesota. It includes more than 2,500 stores (e.g., Cub Foods, Albertsons, Jewel-Osco, Shoppers) with a range of formats (value stores, superstores, full service) and revenues over US$37 billion. SUPERVALU also operates supply chain services to independent retailers (http://www.supervalu.com).

SUPERVALU operates one of the largest grocery supply chain networks in the United States with more than 20 million square feet of distribution facilities located to support their own retail operations and other customers. These facilities manage the delivery logistics of products from hundreds of manufacturers to approximately 5,000 retail end points.

SUPERVALU has worked to minimize its environmental footprint. Initiatives to achieve this range from promoting customer involvement in the company's recycling programs to reducing energy usage in stores and offices SUPERVALU has an environmental stewardship council, ValuEarth^{TM}, that studies, recommends, and oversees ways in which the company can minimize its environmental footprint.

SUPERVALU recently joined the EPA GreenChill Advanced Refrigeration Partnership Program. As part of this program, SUPERVALU now uses only non-ozone-depleting refrigerants in all commercial refrigeration applications in new construction and store remodels involving rack additions or replacements; reduces emissions of ozone-depleting and greenhouse gas refrigerants every year; and participates in an industry/government research initiative to assess the performance of advanced refrigeration technologies.

SUPERVALU has an active waste reduction program that includes recycling and composting. For example, SUPERVALU converts collected plastic bags into composite lumber for benches, and 429 million beverage containers were returned to SUPERVALU stores as part of facilitation-in-state rebate programs (http://www.supervalu.com).

Waste reduction is further achieved in select stores that have a produce extractor to remove liquid waste from the produce waste stream, which reduces the volume and weight of waste sent to landfills Composting is also done in select stores. Albertsons stores in Seattle turned their store compactors into compost containers, which are fille and removed from the stores weekly, and Acme stores in Pennsylvania fil compost totes that are collected on a daily basis and are converted into high-grade soil. Select Shaw's stores in the Northeast have partnerships with farmers to utilize store produce waste for animal feed. Reusable bags are also sold in stores to reduce plastic use, with nearly 2 million bags purchased during 2007 (http://www.supervalu.com).

Between 2000 and 2007, SUPERVALU invested US$95.2 million in capital on energy projects, which saved over 340 million kilowatts of energy per year (http://www.supervalu.com). All its new stores are designed with low-energy white thermoplastic polyolefin roofin systems. Energy-efficien lighting is increasingly used: natural light, with skylights; LED signs and lighting; and in select facilities, roof-mounted solar power collection systems for green roofs. SUPERVALU designed and built its firs Leadership in Energy and Environmental Design (LEED) store in Worcester, Massachusetts. The store opened in January 2005 and was the second LEED-certifie grocery store in the country.

Whole Foods Market, Inc.

Whole Foods Market, Inc. opened its firs store in 1980, with sustainability as the key driving force, when three local businessmen decided the natural foods industry was ready for a supermarket format. Currently the company includes 270 stores in North America and the United Kingdom (www.wholefoodsmarket.com). The company's three main environmental strategies are to support sustainable agriculture/organically and biodynamically grown foods in order to reduce pesticide use and promote soil conservation; reduce waste and consumption of nonrenewable resources through recycling, reusable packaging, reduced packaging, and water and energy conservation; and encourage environmentally sound cleaning and store maintenance programs.

Since the beginning, Whole Foods Market has supported organic agriculture, when the organic network was difficul to source and distribute. As a result, they set up their own produce distribution company in California and this evolved to the company getting involved in the packaging, storage, and shipping procedures to ensure the quality of the products from the farms to the stores. This commitment to organic sourcing continues as they serve as the sole retail representative on the National Organic Standards Board to help in the development of the regulations on these products in the United States. They also work with ranchers and poultry producers to develop hormone and antibiotic-free alternatives for their customers. They have developed quality standards and work with manufacturers to provide products that meet the standards. Further, a significan part of their effort in sustainable food has included educating their customers.

Whole Foods Market further reduces its reliance on nonrenewable resources through purchases of renewable energy and reducing its own carbon footprint. Some of the carbon reductions include fueling trucks with biodiesel at four of the Company's nine distribution centers. Additional efforts to reduce waste include composting of food waste at stores and facilities, recycling materials like plastics and glass and aluminum at stores, and selling canvas and reusable shopping bags. In their stores to reduce resource use, building construction has focused on waste reduction and use of renewable materials. Further, they use less toxic cleaning products in their buildings and educate customers about the positive impact that can be made in air and water quality by using these alternative products.

Whole Foods Market leads in its customer education. Every store has a Take Action Center that provides a wide variety of information on local, regional, national, and international issues such as genetic engineering, organic foods, pesticides, and sustainable agriculture, legislation, and the tools customers need to effectively participate in shaping those issues. Wholefoodsmarket.com then provides detailed information about these issues.

Marks and Spencer

Marks and Spencer is a leading retailer in the United Kingdom, with over 600 stores located throughout the United Kingdom, and also has 240 stores worldwide, including over 219 franchise businesses, operating in 34 countries. With a range of sustainability activities, according to a survey in 2007, Marks and Spencer is considered to be making the most genuine green efforts of all companies in the United Kingdom (Chatsworth Communications, 2007).

Marks and Spencer conducted research which identifie indirect impacts as a key focus area, since production and use of products were around 12 times greater than those from their operations. "We believe it is responsible to try to balance the need for sufficien quantities of high-quality foods with meeting Environmental concerns" (Marks and Spencer, http://www.marksandspencer.com).

In 2003, Marks and Spencer launched a set of standards to cover the management of their supply chain for fruit, vegetables, and salads. This was done in collaboration with suppliers, government bodies, and other organizations and covers aspects of production from "fiel to fork." The standards include traceability, minimizing pesticide use, ethical trading, support for non-genetically modifie foods, and food safety. Further, Marks and Spencer is sponsoring research on food miles to understand the role that issue plays in their supply chain.

With this "farm-to-fork" strategy, the company has committed to organic production. However, since organic foods are typically more expensive, Marks and Spencer has modifie their margin for these products to keep them as affordable options for their customers by making the same cash profi margin on organic as equivalent conventional products, rather than a straight percentage.

For fish Marks and Spencer has been helping to ensure protection of fis stocks by avoiding the purchase of fis where the origin of the catch is unknown. They are also working with the Marine Stewardship Council. and developed their own code of practice for fis farms.

For direct impacts, Marks and Spencer is aiming to be more efficien in a number of areas. For energy usage, they monitor electricity in most stores and have been able to readily identify inefficiencie such as air-conditioning systems that have been left on accidentally. Marks and Spencer is replacing ozone-depleting refrigerant gases with less harmful hydrofluorocarbon (HFCs), and new refrigeration systems are designed to use minimum levels of HFCs and to consume around 4% less electricity than the previous units. They also operate a number of secondary refrigeration systems that distribute cold temperature to the main refrigerators through a network of pipes fille with glycol.

For transport, they have been findin ways to make road transport more efficien and diverting as much transport from the road to other more efficien options such as rail transport. The company has also established water efficien y practices, with stores fitte with infrared sensors to reduce the number of automatic flushe in restrooms, and the company is installing gray water recirculation systems that reuse water for flushin toilets.

For waste, the company found that nearly half of all packaging was used to protect products as they were transported. To reduce the amount of this type of packaging, they have been using reusable plastic crates to transport fresh food products. These crates replaced over 25,000 tonnes of cardboard and many of the original reusable plastic crates were in service for over 20 years before they wore out. Finally, they donate unsold food to local charities, donating the equivalent of a quarter of a million carrier bags of unsold food, or about 1,200 tonnes every year.

Conclusions

The direct impacts food retailers have on the sustainability of the supply chain are in energy use and refrigeration, waste management, and distribution of the operation. As was noted by the examples, however, the greatest impact the retailer has is on the indirect impacts through their supply chain—namely, the products offered in the stores. The examples also demonstrated that there are opportunities for more sustainable practices to improve such impacts by providing sustainable food choices and educating their customers about why those choices are preferred.

References

BBMG. 2007. *Conscious Consumers Are Changing the Rules of Marketing. Are You Ready?* Highlight from the BBMG Conscious Consumer Report. November 2007.

Braschkat, J., A. Patyk, M. Quirin, and G.A. Reinhardt. 2003. "Life cycle assessment of bread production—a comparison of eight different scenarios." In: *Life Cycle Assessment in the Agri-Food Sector: Proceedings from the 4th International conference, October 6–8*, Bygholm, Denmark.

Chatsworth Communications. 2007. *UK Big Business Puts Image and Consumer Pressure Ahead of Genuine Concern for the Environment, According to First Major UK Survey of Opinion Formers.* Press Release. September 23, 2007.

Davies, T., and D. Konisky. 2000. *Environmental Implications of the Foodservice and Food Retail Industries.* Discussion Paper 00-11. Resources for the Future. Washington, DC.

DOE (United States Department of Energy). 2006. *Commercial Energy Use.* Available from http://www.eia.doe.gov/kids/energyfacts/uses/commercial.html. Accessed September 19, 2008.

Franklin Associates. 1995. *Grocery Packaging in Municipal Solid Waste.* 1995 Update for Grocery Manufacturers of America.

Greenbiz.com. 2008. *Wal-Mart Expands Sustainability Efforts with Coffee, Trucks.* Available from http://www.greenbiz.com/news/2008/04/02/wal-mart-expands-sustainability-efforts-with-coffee-trucks. Accessed April 2, 2008.

Keoleian G.A., J.W. Bulkley, R. DeYoung, A. Duncan, E. McLaughlin, D. Menerey, M. Monroe, and T. Swenson 1991. *Waste Reduction in Food Retail: Case Study Report of the People's Food Co-op.* Available from http://css.snre.umich.edu/main.php?control=detail_pub&pu_report_id=45. Accessed September 19, 2008.

Landor. 2007. *A Time For Green Brands: A View from the 2007.* ImagePower® Green Brands Survey. Available from http://www.wpp.com/NR/rdonlyres/EC7958CA-D207-4BF2-BC9E-4474199C789B/0/thestore_005_LandorUK_ATimeforGreenBrands_PhilGandy.pdf. Accessed September 19, 2008.

McTaggart, J. 2008. Best of both worlds. *Progressive Grocer*, April 1, 2008.

Moore, B. 2008. Look below: It's what's under those cartons that can help improve sustainability. *Progressive Grocer*, April 1, 2008.

Opinion Leaders. 2007. *Supermarkets Thematic Review: Stakeholder Consultation Report.* Available from URL http://www.sd-commission.org.uk/publications/downloads/supermarkets_stakeholder_consultation_report.pdf. Accessed September 19, 2008.

Safeway. 1999. *1998 Annual Report.* Pleasanton, California.

Stop and Shop. 2008. *EPA Recognizes Stop and Shop and Giant Food; Only Supermarkets Receiving 2007 ENERGY STAR Leaders Recognition.* Press Release, February 7, 2008.

Sustainable Food News. 2007. *Survey: Consumers Trust Activists, Grocers for Food Info.* December 6.

Wegmans. 2008. *Mary Ellen Burris Column: The S (Sustainability) Word.* Available from http://www.wegmans.com/webapp/wcs/stores/servlet/MEBDetailView? langId=-1&storeId=10052&catalogId=10002&productId=566563. Accessed September, 19, 2008.

Chapter 10

Sustainability in Food Service

John Turenne

Introduction

More and more food services are shifting from conventional food service, focus on quantity and convenience over quality and sustenance often with highly processed, heat-and-serve food that lacks nutritional value, to more sustainable food programs that provide food that supports the well-being of the population and planet. While much of these efforts have been driven by cost reduction, there is also a growing consumer demand. In 2007, 62% of consumers said that they would choose to eat at a restaurant they thought was environmentally friendly (NRA, 2007). There are also many operations that have incorporated sustainability strategies in their philosophy, and are leading changes. For example, Yale University was one of the firs institutional food service providers to develop a menu based on seasonal and locally produced food. Chipotle continues to challenge the concept of quick service with fresh preparation, sustainable food selection (e.g., grass-fed beef), and sustainable building construction.

Food service operations range from fast food chains, independent restaurants, hotel/large volume restaurants, and institutional cafeterias like universities and hospitals. The Food Service industry is one of the largest employers in the United States, employing about 13 million individuals (NRA, 2007). Regardless of the scope of this industry, in many ways, food service operations are not much different from day-to-day means of feeding a family. The ingredients in the food need to be grown or raised and then to be packaged and stored before it gets prepared and then eaten. Food service operations simply take these aspects of supply chain and produce it in volume. This chapter reviews the main steps of food service operations and how they can implement more sustainable approaches for:

1. Procurement
2. Storage
3. Preparation
4. Service

Procurement

As discussed in previous chapters, agricultural production is the primary source for sustainable improvements in the food system due to the significan environmental impacts at this stage. As a result, the sourcing from more sustainable food and beverage options should be the main focus for improving the impact of food service operations. When working in a sustainable food service operation, some considerations include; findin food and ingredients grown organically or with integrated pest management (IPM) or with the least negative impact to the environment, where the food originates and the distance the food traveled, or if the food is certifie by any sustainable programs, such as the Marine Stewardship Council or Fairtrade.

As an example, Bon Appétit, a United States food service provider, evaluated their sourcing from a life cycle perspective and define several steps to reduce their impact (Bon Appétit, 2007):

- Reduce the use of beef by 25%—livestock production is responsible for 18% of greenhouse gas emissions.
- Source all meat and poultry from North America—80% of the energy used by the food system comes not from growing food, but from transporting and processing it.
- Source nearly all fruits and vegetables from North America, using seasonal local produce as a firs preference and using tropical fruits only as "special occasion" ingredients—most bananas have traveled 3,000 miles in high-speed refrigerated ships to reach an American breakfast plate. A local apple might be grown within 10 miles.

Ordering and Distributors

The industry has moved to menus that are more complex and diverse, with concern for the origin and make-up of food most typically considered only when it relates to cost, contrary to the needs stated above for sustainable sourcing. In more conventional models, the need for expediency and cost control results in as few distributors as possible. *One-stop shopping* is a term often used to control the number of orders that must be placed, deliveries received, and invoices processed. As a result, distribution is one of the biggest challenges faced in sourcing more sustainable food options.

Although there are cost and efficien y benefit to working through a distributor as just mentioned, there are many other reasons to try and deal directly with producers/farmers. For one, the simple face-to-face interaction with the person who is responsible for producing the food goes a long way with staff development. There are some considerations when working with smaller and local producers. Packaging capacities may be different for these farmers and producers than for larger, more industrial producers. For example, fresh picked salad greens may be bagged in loose plastic bags rather than the sealed 3 lb prewashed packages. Not all producers have the resources to make the necessary deliveries. Another concern often raised is the issue of liability insurance required by the institution to prevent possible litigation from customers, should an unfortunate outbreak of a food-borne illness occur. In reality, this is an unfortunate misconception of where the real dangers lie in food handling. The smaller, local farmers and producers, in fact, can be more easily monitored and have much more at stake in their livelihoods than to risk the danger of not taking safe steps in food handling.

One way around having to make so many direct connections with producers or farmers on a day-to-day basis is to work through small, local distributors. Often, there are existing vendors in a geographic area, who can act as a "middle man." They have existing relationships (or can develop relationships) with the producers/farmers and act as the clearinghouse for the food service to order food from. Therefore, there is less need for individualized and time-consuming ordering, receiving, etc. This model can also be helpful in overcoming liability insurance requirements, as the distributor carries this coverage. However, it is important to note that in order to make this a viable model for all parties—producer/farmer, distributor, and food service operation—everyone must be up-front and open about costs and payments. This will not work if the distributor underpays what the producer/farmer deserves to be paid.

To make sense of the sourcing and distributing options available, it is important for an operation to set its goals and fin the best means to achieve those goals. For example, the Yale Sustainable Food Project, founded in 2001 by Yale students, faculty, and staff, President Richard Levin, and chef Alice Waters, directs a sustainable dining program at Yale University. The emphasis for the program is sourcing food from local and sustainable options. To meet these goals for procurement, they have worked with a range of distributors and farmers and, as a result, developed a set of guidelines for sourcing as follows (Yale Sustainable Food Project):

Vegetable guidelines
First tier (ranked in order of preference)

- Connecticut organic
- Connecticut ecologically grown
- Regional organic
- Regional ecologically grown
- Connecticut conventional (small-scale operation)
- Regional conventional (small-scale operation)

Second tier (ranked in order of preference)

- Connecticut conventional (medium-scale operation)
- Regional conventional (medium-scale operation)
- U.S. organic (small-scale operation)
- Connecticut conventional (large-scale operation)
- Regional conventional (large-scale operation)
- U.S. ecologically grown (small-scale operation)

Third tier (ranked in order of preference)

- U.S. organic (medium/large-scale operation)
- North America organic
- U.S. ecologically grown (medium/large-scale operation)
- International organic
- U.S. conventional (small-scale operation)

Fruit guidelines

First tier (ranked in order of preference)

- Connecticut organic
- Connecticut IPM
- Regional organic
- Regional IPM
- Connecticut conventional (small-scale operation)
- Regional conventional (small-scale operation)
- Connecticut conventional (medium-scale operation)

Second tier (ranked in order of preference)

- Regional conventional (medium-scale operation)
- U.S. organic (small/medium-scale operation)
- U.S. IMP (small/medium-scale operation)
- Connecticut conventional (large-scale operation)
- U.S. organic (large-scale operation)
- U.S. IPM (large-scale operation)
- International organic
- U.S. conventional

Meat and poultry guidelines

First tier (ranked in order of preference)

- Connecticut free range/pasture fed
- Connecticut organic
- Regional free range/pasture fed
- Regional organic
- Regional conventional (small-scale operation)

Second tier (ranked in order of preference)

- U.S. free range/pasture fed
- U.S. organic (small/medium-scale operation)
- Conventional (small/medium-scale operation)
- U.S. organic (large-scale operation)
- U.S. conventional (large-scale operation)

Menu Planning

As mentioned above, the industry trend has been quantity and convenience. Menus are based on high-impact processed foods, frozen foods, and global distribution. This has also resulted in little, if any, preparation and skill to handle, therefore, minimal staff. While customers have become used to having lots and lots of choices, they have lost much in the way of good, quality food, and at the same time have more detrimental food choices. A recent study in the United Kingdom found that expensive meals were more environmentally preferable compared to cheaper meals, which have a higher carbon footprint (University of Nottingham, 2008).

Sustainable food can encompass any type of menu and concept if incorporated correctly. Customers should be able to enjoy both traditional and seasonal specialty items like gourmet pizzas, fresh sandwiches, salads, and grilled items. Time and resources need to be invested to understanding what to offer as a menu. The food should be carefully prepared in ways that maintain and respect purity, freshness, and fl vor, considering staff talents and capabilities.

A successful practice in sustainable programs has been to scale back on the amount of choices offered and focus more effort on fewer options, in other words, practice a "quality over quantity" mentality. By doing so, the staff's time is focused on handling the whole, fresher food rather than on using their time on so many stations and options. This may include more preparation at the point of service. By adding this point of interaction with the staff member, the customer will be more excited and will no longer think the food comes from some mystery place behind the kitchen walls. This limited menu approach also limits overproduction, a key source of waste. Chipotle Mexican Grill, a fast casual chain in the United States, has had success with this practice, having only fi e types of menu items (with several variations) prepared in front of the customer from fresh produce prepared on-site.

Often one of the most misguided reasons that people say an institutional sustainable food program cannot be successful is that they do not have the right facilities, kitchen, or equipment. Although attention must be given to these factors, menus can be developed accordingly. Take storage and work stations into account when implementing every menu and recipe. Is there enough shelf space or refrigeration to hold cases of whole cabbage, instead of preshredded? Or is there enough room and shelving in a dry goods area to properly store fresh tomatoes, potatoes, and winter squash?

An additional step to developing a menu that is based on more sustainable food is to understand the seasonal availability of food and leverage the possibilities of locally sourced food. While a menu typically is not composed solely of local products, eliminating the use of blatantly out-of-season foods will reduce their impact and provide fresh ingredients for quality menu items. Seasonal food charts are available from state's departments of agriculture or extension programs.

Burgerville (www.burgerville.com), a quick service chain in the Northwest United States, purchases food locally and provides seasonal ingredients in its salads and limited seasonal offerings such as strawberry milkshakes, strawberry smoothies, and strawberry shortcake in the spring.

The Yale Sustainable Food Project, described previously, has implemented their purchasing guidelines with the following sample menus, with the focus on seasonality and local sourcing (Yale Sustainable Food Project):

Fall
Roasted corn and tomato soup
Watermelon, arugula, and ricotta salata salad
Lamb and feta patties with cucumber tzatziki
Pappardelle with ricotta and sautéed Swiss chard
Roasted heirloom and fresh tomato pizza
Chive and parsley mashed potatoes
Roasted zucchini and yellow squash
Apple crisp

Late fall
Butternut squash soup
Fennel and Parmesan salad
Roast pork loin with cranberry compote
Leek and potato galette
Roasted squash, maple syrup, and sage pizza
Roasted romanesco caulifl wer
Cranberry oatmeal cookies

Winter
Boston clam chowder

Spinach salad with toasted almonds
Whole roast chicken with herbs
Roast chuck eye
Pizza with potatoes and rosemary
Parsnip and potato purée
Squash gratin
Brownies

Spring
Asparagus soup
Mixed greens with roasted shallot vinaigrette
Grass-fed beef hamburgers
Pappardelle with asparagus lemon sauce
Asparagus and Parmesan pizza
Olive oil and garlic mashed potatoes
Sautéed spinach with garlic
Ricotta tart

Storage

Once the food makes it to the facility, it must be stored. The main impacts during this stage of the operation come from waste and refrigeration.

Solid waste generation is significan in food service operations and is composed of packaging and food—with packaging being the predominate source of solid waste. This is because, in conventional food services, packaging plays a significan part in the control needed to operate large food service operations, in addition to performing the function of delivering safe and quality products. Consistency in package sizes, taste, and quality have been the backbone to long-term success in managing costs, staff handling, and customer expectations. Packaging is primarily from cardboard and paper, which is recyclable. Reusable packages can be used such as boxes, baskets, and egg crates. Additional packaging should and can be minimized. As a result, the total amount of packaging waste going to the landfil can be minimized.

Food service and food retail operations represent the largest commercial users of refrigeration, 23.6 and 39.0%, respectively (DOE, 2006). Refrigeration requires energy, and may use chlorofluorocarbon (CFC). CFCs are known ozone-depleting substances and the United States

Environmental Protection Agency regulations issued under Sections 601-607 of the Clean Air Act have phased out the production and import of ozone-depleting substances, consistent with the schedule developed under the Montreal Protocol. However, many operations may have CFC refrigeration in older units. CFCs should be recycled and replaced with more environmentally preferable options.

Preparation and Service

The main impacts of operations during preparation and service are energy, water, and waste. Food service operations comprise 7% of the all commercial building energy use in the United States (DOE, 2006). According to the Environmental Law and Policy Center, cooking equipment consumes the largest share of energy in most restaurants (35%). This is followed by heating and cooling systems (28%), dishwashing (18%), lighting (13%), and refrigeration (6%). The most effective way to reduce such energy usage is to plan cooking of items to minimize equipment on-time (such as items that can be cooked in bulk once or twice a day and the ice machine could be run during off-peak hours, overnight) and to turn off idle equipment. In addition, all equipment should be regularly inspected, cleaned, and maintained. Further energy savings can be found by using Energy Star appliances like fryers, steamers, reach in refrigerators or freezers, holding cabinets, and water heaters. In addition, hot water should be around 140°F at the faucet of the pot sink closest to the dish machine (Energy Star, 2007).

Kitchen exhaust hoods are intended to capture and contain the grease and heat produced in the kitchen, when they are not working correctly extra heat ends up in the kitchen (Food Service Technology Center, 2007). Maintenance and cleaning of the ducts and fans will help reduce energy requirements. Well-engineered and properly installed UL-listed (Underwriters Laboratories) hoods reduce required exhaust volume, reduce heating/cooling costs, and reduce fan energy use (Food Service Technology Center, 2007). Additional considerations for exhausting include use demand-based ventilation; use a wall-mounted canopy instead of an island canopy hood; use a proximity hood over light-duty appliances (griddles, fryers); group like-duty appliances under the hood; push equipment as far back under the hood as possible; install side panels and a front lip; keep makeup air delivered near the hood at low velocity; and use evaporative cooling and direct-fire heaters on makeup air (Food Service Technology Center, 2007).

For climate control, programmable thermostats or sensory-based thermostats with "night setback" modes should be used. There is Energy Star-qualifie heating and cooling commercial equipment, which uses 7–10% less energy than standard equipment (Energy Star, 2007). Using ceiling fans can reduce air-conditioning use and total energy use; energy use falls by 4–5% for every degree that you raise your cooling thermostat (Energy Star, 2007).

Compact fluorescen lamps (CFLs) and light-emitting diodes (LEDs) should be used instead of incandescent bulbs. CFLs can be used in the back and front of the house, where LEDs are more typically used in the front of the house. In addition, occupancy sensors can be used in closets, storage rooms, break rooms, restrooms, and even walk-in refrigerators and turn off all possible lighting during the daylight (outdoor and some indoor locations).

It has been estimated that water used in restaurants is mostly from restroom water use—ranging from 50% in full-service restaurants to 80% in fast food restaurants (DPPEA 2, 1999). So use of low-fl w faucets and/or automatic turn-off faucets, and low-fl w toilets in restrooms are good options. Additionally, water use can be reduced with dry cleanup procedures and will reduce the amount of food waste that enters the drains and, thus, disposal of food waste via the sewer system.

From the preparation and service aspect, waste comes from food and disposable products. According to the Environmental Law and Policy Center, organic waste constitutes 24–50% of a restaurant's waste stream. This organic waste includes noncontaminated edibles, food scraps, and waste oils. The most common reason for food waste is overproduction, comprising 63% of the food waste (Food Management, 2007). Some food wastes can degrade in a landfill but do so at a very low rate of 25–50% over 10–15 years, with the remainder of the food waste found to retain its original weight, volume, and form (DPPEA 1, 2000). Given that is it near impossible to prepare exactly the amount of food needed, leftover food and scraps should be separated and discarded for composting (rather than discarded together as one in trash receptacles that are emptied in landfill or burned in incinerators). This separation is easily achieved, especially during preparation. As a result this practice is becoming more widespread. Plate waste on the other hand is more challenging. At the very least food materials should be scraped off, so they are not discharged to a wastewater treatment plant to contribute to increased levels of BOD (biological oxygen demand), COD (chemical oxygen demand), TSS (total suspended solids), and FOG (fats, oils, and grease). Cooking oil can be collected and is increasingly being used for biodiesel conversion for use in vehicles.

The use of disposable service ware has increased dramatically. When dealing with service ware, utilization of reusable plates minimizes waste. In addition, disposable table coverings should not be used, rather use of reusable table linens, or no table covering at all, are preferred. Polystyrene had been commonly used for food containers. It is derived from benzene, a carcinogen and ozone depleter. As a result, polystyrene has been banned in some municipalities. If this approach is impractical, then minimizing resources used and total waste from these materials can be done—by using products with recycled content (e.g., napkins and containers) and recyclable products (e.g., bags and beverage containers). Sodexo Inc., an international food service provider, uses recycled content napkins and has calculated that change to save more than 23,000 trees, nearly 10 million gallons of water, 5.5 million kilowatts of energy, 5,000,000 gallons of oil, kept 41 tons of pollutants out of the air, and kept 4,131 cubic yards of paper out of landfill in just 1 year (Sodexo, 2008). Sodexo is also introducing dispensers that allow only one napkin to be dispensed at a time. In preliminary use, the dispensers have reduced paper use by 25–50% at some sites (Sodexo, 2008).

Food safety is an additional consideration for operations. It has been estimated that 40% of all reported food-borne illnesses were traced to food eaten at restaurants (CDC, 1996). This is commonly caused by mishandling of food during preparation. An effective hazard analysis and critical control point (HACCP) system could prevent much of these problems.

Implementing Sustainable Operations

Food purchases, menus, preparation methods, equipment use, water use, recycling, and composting are the main considerations for implementing a sustainable food service program. However, actually making it happen, for long-term success requires an internal champion and staff commitment. The staff will need to know *why*, as well as *how* to go about executing the sustainability program.

First of all, there needs to be a champion within the ranks of leadership. This person must already demand respect and authority, and will be willing to go to extremes to lead, support, and nurture the rest of the staff. He or she need not be completely experienced in the aspects of sustainable food, but must be willing to learn and subscribe to the principles.

After this champion, it is equally important to devote time, energy, and resources into all the staff. There are two ways that staff should be trained. One is to teach them how to go about this new style of service and menu, and the second and probably more importantly is to explain why it is being done so they are brought into the program and do it even if the champion or leaders are not around. Ways that have been successful in this focus have been visits and conversations with producers and farmers; taste tests; sustainable media events; roundtable discussions; as well as guest chefs and speakers; or something as simple as sitting down to a lunch.

To train how to undergo a sustainable food program, institutional leaders must invest time and resources. Use time outside of normal service to conduct hands-on training with whole fresh foods. It is not uncommon for staff to have forgotten how to handle fresh food, or they may never have had the skills in the firs place. It may take time to nurture and teach team members so as to ensure the program's success.

To complete the implementation, the operation has to reach to its customer. Education and marketing are important to ensure that the customer is aware and included in the program. This has been done by inviting farmers to lunch, staff discussions with customers, providing informative signs/posters, and establishing a group of regular customers to be spokespeople for their peers is one way. Bon Appétit, previously mentioned, has signs, fl ers, and a web site dedicated to educating its customers about the food sourcing changes they have made. Their web site also includes a dietary/environmental impact tool to extend the sustainability experience to the food choices made at home.

Case Study: Pizza Fusion

The new United States pizza chain, Pizza Fusion (www.pizzafusion .com), will be used to demonstrate sustainable implementation across all the aspects discussed in this chapter. Pizza Fusion begins its sustainable execution with a focused menu of pizza, sandwiches, and salad, using natural and organic ingredients. The menu emphasizes quality ingredients and taste (e.g., New York strip steak). To provide these ingredients they have established sourcing partnerships with key suppliers, such as Applegate Farms for preservative-free, antibiotic-free pepperoni and prosciutto and Crystal's for organic sugar, the only organic sugar farm in the United States.

Waste and resources at Pizza Fusion are reduced with in-house recycling and use of post-consumer content disposable products. Energy consumption and impacts are minimized by using company hybrid vehicles for delivery, recycling heat from ovens to warm the restaurant and water, use of CFLs and occupancy sensors, use of Energy Star appliances, and remaining power usage offset with the purchase renewable energy credits. Water usage is reduced by one-third, with recycling of water from sinks into toilets. Also, facility maintenance is done with environmentally preferable cleaners, organic cotton is used for uniforms and promotional items, and restaurant construction is done with renewable, recycled, and reused products.

Finally, Pizza Fusion completes its execution by reaching out to its customers and providing educational resources. For example, each store offers free cooking classes to children that focus on organic foods and the importance of sustainable living. Each store also provides a discount to customers to bring back pizza boxes for recycling.

Conclusions

Incorporating or even transforming to a sustainable food service operation requires more than a few light bulb changes and recycling of cardboard; it includes the realization that sustainable food services have greater impacts, especially from the food served. This is because food service operators are responsible for significan volumes of food moving through the food system each year. Responsible practices of sustainable food purchases, careful menu planning, waste reduction, composting food waste, recycling solid waste, energy reduction, water reduction, and education to staff and customers can be incorporated into any operation. Ultimately, sustainable food services promote environmental, economic, and social well-being.

References

Bon Appétit. 2007. *An Inconvenient Tooth: Food Is Major Contributor to Climate Change: New "Low Carbon Diet" Aims to Take Bite Out of Global Warming.* Palo Alto, CA. Available from http://www.bamco.com/PressRoom/press-pre-041707.htm. Accessed April 17, 2007.

CDC. 1996. Surveillance for foodborne-disease outbreaks—United States, 1988–1992. *Morb. Mortal. Wkly Rep.* 45(SS-5):1–66.

DOE (United States Department of Energy). 2006. *Commercial Energy Use.* Available from http://www.eia.doe.gov/kids/energyfacts/uses/commercial.html#TYPES. Accessed September 19, 2008.

DPPEA 1 (North Carolina Department of Environment and Natural Resources Division of Pollution Prevention and Environmental Assistance). 2000. *A Fact Sheet for Licensed Garbage Feeders.* Available from http://www.p2pays.org/ref/04/03991.pdf. Accessed September 19, 2008.

DPPEA 2 (North Carolina Department of Environment and Natural Resources Division of Pollution Prevention and Environmental Assistance). 1999. *A Fact Sheet for Managing Food Materials.* Available from http://www.p2pays.org/ref/03/02792.pdf. Accessed September 19, 2008.

Energy Star. 2007. *Putting Energy into Profits: ENERGY STAR Guide for Restaurants.* Available from http://greenrestaurants.org/documents/Energy_Star_Restaurants_Guide.pdf. Accessed September 19, 2008.

Environmental Law and Policy Center. *Going Greener Guide.* Available from http://www.greenrestaurants.org/documents/GoingGreenerGuide.pdf. Accessed September 19, 2008.

Food Management. 2007. *A Plate Is a Terrible Thing to Waste.* Available from http://food-management.com/ar/fm_imp_16651/. Accessed September 19, 2008.

Food Service Technology Center. 2007. *The Energy Efficient Kitchen.* Available from http://www.fishnick.com/design/eek/kitchen.html Accessed September 19, 2008.

NRA (National Restaurant Association). 2007. *2008 Restaurant Industry Forecast.* Available from http://www.restaurant.org/research/forecast.cfm. Accessed September 19, 2008.

Sodexo. 2008. *Sodexo Marks One Year of Recycle Paper Napkin Program with Significant Environmental Impact Announcement.* Press Release, March 31, 2008.

University of Nottingham. 2008. *Food for Thought.* Press Release, May 19, 2008.

Yale Sustainable Food Project. Available from http://www.yale.edu/sustainable-food/index.html. Accessed September 19, 2008.

Chapter 11

Sustainability Principles and Sustainable Innovation for Food Products

Cheryl Baldwin and Nana T. Wilberforce

Introduction

Sustainable innovation and product development is a process of designing environmentally and socially preferable, commercially successful products. The process is life cycle-based and considers sourcing of materials, production, packaging, delivery, and the end-of-life of the product. A business that effectively develops sustainable products integrates sustainability considerations into the business as a whole. Such businesses have seen real value in differentiation from and outperforming competitors (through such means as cost reduction, reduced risks and, improved reputation) (Business Link, 2007). Further, this is an effective means for a company to contribute to global improvements in areas such as climate change, water scarcity, and malnutrition.

This approach has been considered by other industries for some time. A result of this work has been the development of Green Engineering. The United States Environmental Protection Agency (EPA, 2007) has define Green Engineering as:

The design, commercialization and use of processes and products, which are feasible and economical while minimizing (1) generation of pollution at the source and (2) risk to human health and the environment. Green engineering embraces the concept that decisions to protect human health and the environment can have the greatest impact and cost effectiveness when applied early to the design and development phase of a process or product.

There are 12 principles of Green Engineering to provide a framework to create and assess the elements of design relevant to maximizing sustainability (Zimmerman):

- Principle 1—Designers need to strive to ensure that all material and energy inputs and outputs are as inherently nonhazardous as possible.
- Principle 2—It is better to prevent waste than to treat or clean up waste after it is formed.
- Principle 3—Separation and purificatio operations should be a component of the design framework.
- Principle 4—System components should be designed to maximize mass, energy, and temporal efficien y.
- Principle 5—System components should be output pulled rather than input pushed through the use of energy and materials.
- Principle 6—Embedded entropy and complexity must be viewed as an investment when making design choices on recycle, reuse, or beneficia disposition.
- Principle 7—Targeted durability, not immortality, should be a design goal.
- Principle 8—Design for unnecessary capacity or capability should be considered a design fl w. This includes engineering "one size fit all" solutions.
- Principle 9—Multicomponent products should strive for material unificatio to promote disassembly and value retention (minimize material diversity).
- Principle 10—Design of processes and systems must include integration of interconnectivity with available energy and materials fl ws.
- Principle 11—Performance metrics include designing for performance in commercial "afterlife."
- Principle 12—Design should be based on renewable and readily available inputs throughout the life cycle.

Green Chemistry is a similar approach, related more specific y to chemical use (and related processes). It aims to reduce waste, eliminate end-of-the-pipe treatment, result in safer products, and have reduced use of energy and resources (EPA, 2008). The result of these approaches, Green Engineering and Green Chemistry (often collectively called eco-design or cleaner production), is the use of science and technology to ensure an increased quality of life through a life cycle–based design and that materials and energy are used efficient y.

Sustainability Principles for Food Products

While the Green Engineering Principles and Green Chemistry discussed have more application to chemical-based industries, their fundamentals can be applied to food and beverage products. This is best done by considering the life cycle impacts of the supply chain. As was stated throughout the book (and thus only briefl reviewed here), agriculture is generally the largest contributor to the life cycle impact of food products. Downstream considerations have lower impacts, but can range depending on the product. In general, food processing is the next most significan contributor to impacts in the supply chain (Yakovleva, 2005). Packaging impacts tend to be very limited compared to other components of the supply chain, less than 2% for most food products, but can be up to 20% due to energy needed to produce the package or when packaging high water-containing products like beverages (Katajajuuri and Virtanen, 2007). Consumer transport to purchase food can be a significan impact (Braschkat, 2003). And final y, consumer use of the food, when including cooking, can also be a contributor to the life cycle impact (Braschkat, 2003).

Putting these finding together with other the other sustainability concepts discussed, the sustainability principles for food products include the following:

1. Ensure safety of the food supply
2. Provide nutritionally dense food
3. Minimize animal products and ingredients
4. Use lower intensity agricultural products
5. Avoid air transport of food
6. Diversify sourcing
7. Process food with minimal inputs; raw materials, water, and energy
8. Process food with zero waste
9. Minimize total packaging
10. Efficient y deliver food to the consumer

The goal of the food supply is to nourish, and if the food is not safe, it cannot provide for that goal. Thus, ensuring the safety of food is fundamental to a sustainable food supply. As mentioned earlier in this book, this includes food free of microbial contamination as well as food with minimal residual agrichemicals and other potential toxins. The goal of nourishment is also closely linked to the nutritional quality of the food. If the food does not provide nutritional density, overconsumption of calories or undernourishment has been a problem.

Given that animal products have significant y greater life cycle impacts than other products, reduction in their reliance is important. However, there are responsible means of producing animal products; thus, when they are used it is best to fin sources with less-intensive feeding and raising practices (grazing vs grain fed) and those raised humanely.

Further, it has been shown that produce can have as high an impact as animal products when grown in high-intensity operations, such as hot houses. Thus, priority should also be given to low-intensity agricultural products. This includes procurement of fresh vegetables or fruits when (and where) they are in season. The link between local food and a reduced environmental impact, at this point, has not been strong. For example, lower-impact growing practices further away have demonstrated benefit over local high-intensity practices.

Given the last principle, efficien distribution is also important. Air transit of food is the least efficien option, and thus should not be used. In addition, further distribution efficiencie should be sought out.

To provide fair markets and wages, greater choice, and more environmentally preferable options, diversificatio of sourcing from growers and suppliers should be practiced. This social consideration also reduces risk of supply issues.

Processing food with minimal inputs including raw materials, water, and energy will reduce the total impact of food processing. This also includes consideration of the impact of each raw material, findin alternatives when feasible. An aim should be toward using renewable energy, or even processing wastes to produce energy. Further, processing food with zero waste (solid, liquid, and emissions) is achievable and should be practiced.

Packaging helps deliver safe food; however, its material use can be minimized and its waste can be eliminated. Light-weight packaging and recycling are some key development options.

Consumer transport, via car, to get their food is a significan impact of the supply chain. As a result, efficien delivery of food to the consumer could reduce the total impact. This challenges existing retailing and food service approaches and should include greater education efforts to consumers and delivery options.

Sustainable Product Development and Innovation

Sustainable Product Development

Now that sustainability principles for food product development have been outlined, how can they be used to develop products? For this discussion, the technical research and development aspects of the process will be of focus, rather than the market research aspects. The research and development aspects of food product development process follow four primary steps:

1. Concept development and testing
2. Prototyping and bench-top development
3. Process, formula, and package development
4. Commercialization

It has been demonstrated through life cycle research that the impact of a product is primarily determined at the onset of the product's development at the concept stage. Envirowise estimates that around 80% of costs are committed at the concept development stage (Envirowise, 2008). This is because while many details about the product are not determined yet (formula or process), the main raw materials and packaging options are often selected at that time. As a result, knowing the life cycle impact potential of product options should be included as early as concept development. This is best done by having executed some life cycle research for your products. Then you have some understanding that can be used at the concept stage. However, reliance on the above-described sustainability principles for food products may also help, given that they are based on the current life cycle literature. For example, understanding that having a recyclable package is an important consideration for a sustainable product, a concept for a self-heating package (i.e., not recyclable) for soup or beverages may not be desirable.

Considering all the sustainability principles for food products upfront in the concept stage is important. However, several sustainability principles usually rise an important consideration as the goals of the project. Prototyping and bench-top development is the stage at which those priorities are tested to present various means to meeting the goals of the project. The result is that during the formula, process, and package development phase, the priorities are inherently included in the project. Life cycle assessment (LCA) is a useful tool during prototyping and development. This ensures the choices being made not only to meet the project priorities, but also that other principles are not negatively impacted. Finally, in the commercialization step, LCA is important to ensure the sustainability goals are met and maintained. There are different ways to undertake LCA. It is expected that the LCA at each stage of the product development process is different from the other. For example, the most complete inventory and impact assessment would be done at the commercialization stage to develop a benchmark and monitor progress. A more simplifie approach (even qualitative) would be done during prototyping.

To summarize, sustainability principles and priorities fi into the product development process in the following way:

1. Concept development and testing—consider sustainability principles
2. Prototyping and bench-top development—test sustainability priorities
3. Process, formula, and package development—include and optimize sustainability priorities
4. Commercialization—confi m sustainability priorities are met and maintained and sustainability principles in balance

Sustainable Innovation

Including sustainability principles into product development presents significan sources for innovation. These principles can be used for new products, product improvements, line extensions, cost reductions, and other development projects (including process and package designs). Four main paths will be presented to demonstrate how sustainability provides a means for innovation:

1. Innovate around the challenges that the sustainability principles present
2. Identify the overlap between sustainability goals and business goals
3. Consider how nature solves the problems presented
4. Understand the life cycle of the product

Innovating around the Challenges That the Sustainability Principles Present

It has been said that behind every innovation is a problem and someone determined to solve it. The process of solving the problem can be broken

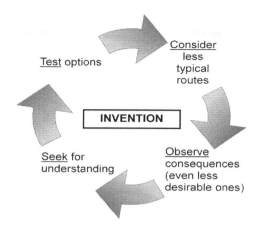

Figure 11.1.   The inventive approach to product development.

down into several steps to maximize the innovative outcome, outlined
in Figure 11.1, the inventive approach to product development. It begins
with definin the problem, and then definin it again with a different
perspective (to seek for understanding), and then moves through testing, and includes exploring difficult-to-sol e problems and less typical
routes. Through such a process and careful observation, unintended consequences may lead to new solutions. However, without a challenge, in
this case the sustainability principles, the innovation may not be found.
This can be done with a single principle, a few, or all at once.

Patent US 7037547 can be used to illustrate this approach with a single
principle. When developing a smoothie beverage, one may assume to
use a dairy source. However, knowing that one of the sustainability
principles of food products is to minimize the use of animal ingredients,
one could use a vegetable protein base like soy. However, soy protein
is known to have off-fl vors associated with it. The patent describes
how exploring a problem, through a less typical route, defl voring the
protein rather than masking the protein fl vor, results in a novel solution.
Further, through careful observation it was found that this solution could
be applied to other protein sources (US 7097874; US 7094439).

Designing to the Overlap between Sustainability Goals
and Business Goals
Bhanti (2008) suggested that business and sustainability goals are often
the same: do more with less (to maximize use of resources), build to last

(for long-term viability), make it scalable (to prepare for change), and design whole networks (since collaboration is key). To look closer at how this approach could help with innovation, when aiming to do more with less, this can be done by findin new uses for stock packaging, idle equipment, or using by-products from a process. For example, many dairy processors pay for whey removal or lose resources in the whey by land spreading the waste. However, whey includes some high-value components that can be used in products. This goes beyond whey protein and includes dairy minerals found in the permeate (the material remaining after the protein is separated out). Separating out the dairy minerals is not a complicated process, and these components could be used to add more health and wellness components to existing products.

To further illustrate, General Mills Inc., a leading producer of baked and unbaked food products in the United States, implemented a project to do more with less by optimizing product design and related production processes to reduce wastes and production costs. This was done with a review of material usage data for a production line that indicated elevated waste generation in the sheeting, makeup table, and packing areas (Minnesota Pollution Control Agency). The main causes were identifie as dough temperature, product and line feasibility (compatibility of equipment with product), employee training and communication, and production line maintenance (Minnesota Pollution Control Agency). A number of processes and line improvements were made including adding guide plates to prevent any product from falling off the edge of the conveyer during packaging; improvements made to the vertical and horizontal guide mechanisms for the cake pans as they are fille with dough and guided toward the ovens; photos contrasting good versus poor quality products were posted beside the production lines to minimize disposal of products misjudged as being unusable; coolant monitoring and maintenance to ensure correct dough temperature; and conveyer belt adjustments scheduled more frequently to improve equipment performance. The result was a 40% reduction (640,000 pounds annually) in ingredient waste on that line alone (Minnesota Pollution Prevention Control Agency). General Mills estimates that the annual plant-wide savings from this program will be nearly US$4.8 million for a reduction of more than 4.2 million pounds of waste (Minnesota Pollution Control Agency). The cost savings come from reductions in the amount of ingredients used, energy used, labor, and waste disposal costs (Minnesota Pollution Control Agency). The company continues to implement a detailed waste-tracking database to provide information from each area of the production line.

Innovation Based on How Nature Solves the Problems

Biomimicry is the concept that nature provides many sources for innovation. This method of innovation has been carefully described by Benyus (1997). The core idea is that nature has already solved many of the problems we encounter. Animals, plants, and microbes have found what works, what is appropriate, and most important, what lasts here on earth (http://www.biomimicryinstitute.org). For example, when considering how to make resealable packages to retain product freshness, one can look to nature to understand where and how nature uses resealable solutions. Surprisingly, nature does not have to rely on adhesive chemicals for resealable situations such as used on the feet of animals like frogs and geckos, where it is a physical design of intermolecular forces (of tiny hairs and pads that produce electrical attractions).

Another example of biomimicry is the three-dimensional logarithmic spiral found in the shells of mollusks, in the spiraling of tidal-washed kelp fronds, and in the shape of our own skin pores through which water vapor escapes (http://www.biomimicryinstitute.org). Liquids and gases fl w centripetally through these geometrically consistent fl w forms with far less friction and more efficien y. PAX Scientifi has designed fans, propellers, impellers, and aerators based on this shape. The technology's streamlining effect has demonstrated reduced energy requirements in fans and other rotors between 10 and 85%, depending on the application; the fan blade design also reduces noise by up to 75% (http://www.biomimicryinstitute.org). The PAX streamlining principle is also used in industrial mixers and water pumps with 1/4 to 1/15 the horsepower typically needed (https://www.paxscientific.com)

Innovation Based on the Life Cycle Assessment

The LCA tool is covered in Chapter 5. LCA can provide insight to the product and process, and even entire system, at many stages. When it is used during product and package design, choices can be compared and processes can be optimized. This is recommended for most sustainable innovation and product development processes to ensure a product assists a reduced environmental impact and avoids the shift of environmental hazards from one area of production, maintenance, or disposal to another. At firs glance many products may seem to be more environmentally preferable than their counterparts; however, it is important to take an LCA approach because while a product may be less harmful in some respects, it may be more harmful in others.

For example, a study by Spitzley et al. (1997) used LCA to evaluate beverage packaging options. The study compared glass, high-density polyethylene (HDPE), paperboard, fl xible pouches, polycarbonate, aseptic carton, polyethylene terephthalate, steel, and composite options across the range of life cycle indicators. They found less energy-intensive materials were the most desirable, usually being lighter weight and refilla le options. Surprisingly, these were plastic options, rather than paperboard. As a result, the feasibility of refilla le packaging options such as refilla le HDPE and polycarbonate bottles was explored. Further, it was found that if refilla le options prove to be less feasible, an alternative approach was found to be a lightweight, fl xible pouch.

Conclusions

The sustainability principles outlined in this chapter for food products pulls together the conclusions from the current state of life cycle research and additional finding noted throughout this book. These principles provide guidance on the key areas to include when evaluating the sustainability of food products and should be used as a source of innovation. The outcome will be more sustainable products, more sustainable businesses, and a more sustainable supply chain that can realize the business and global benefit of such practices.

References

Benyus, J. 1997. *Biomimicry: Innovation Inspired by Nature*. New York, NY: Harper Collins.

Bhanti, C. 2008. *The Future as a Design Challenge*. Available from http://www .greenbiz.com/users/chhaya-bhanti. Accessed March 10, 2008.

Braschkat, J., A. Patyk, M. Quirin, G.A. Reinhardt, 2003. "Life cycle assessment of bread production—a comparison of eight different scenarios." In: *Life Cycle Assessment in the Agri-Food Sector: Proceedings from the 4th International conference, October 6–8*, Bygholm, Denmark.

Business Link. *Make Your Business More Sustainable.* 2007. Available from http://www.businesslink.gov.uk/bdotg/action/detail?r.l1=1079068363&r.l3=10794 04746&type=RESOURCES&itemId=1079404840&r.l2=1079363670&r.s=sc. Accessed September 19, 2008.

Envirowise. 2008. *Packaging Design for the Environment: Reducing Costs and Quantities.* Available from http://www.envirowise.gov.uk/GG360. Accessed September 19, 2008.

EPA (United States Environmental Protection Agency). 2007. *Green Engineering.* Available from http://www.epa.gov/oppt/greenengineering. Accessed September 19, 2008.

EPA (United States Environmental Protection Agency). 2008. *Green Chemistry.* Available from http://www.epa.gov/greenchemistry. Accessed August 19, 2008.

Katajajuuri, J.-M., and Y. Virtanen. 2007. "Environmental impacts of product packaging in Finnish food production chains." In: *Proceedings from the 5th International Conference on LCA in Foods, April 25–26, 2007,* Gothenberg, Sweden.

Minnesota Pollution Control Agency. *General Mills Inc., Chanhassen Bakeries and Foodservice Plant: A Case Study.* Available from http://proteus.pca.state.mn.us/oea/publications/dfe-generalmills.pdf. Accessed September 19, 2008.

Spitzley, D.V., G.A. Keoleian, and J.S. McDaniel. 1997. *Project Summary: Life Cycle Design of Milk and Juice Packaging.* Available from http://css.snre.umich.edu/css_doc/CSS97-08.pdf. Accessed September 19, 2008.

US 7037547. Akashe, A. and R. Meibach. Method of defl voring soy-derived materials for use in beverages. May 2, 2006.

US 7094439. Akashe, A., C. Jackson, A. Cudia, and J. Wisler. Meibach. Method of defl voring whey protein. August 22, 2006.

US 7097874. Meibach, R., A. Akashe, G.W. Haas, and L.G. West. Process for debittering peanut hearts. August 29, 2006.

Yakovleva, N. 2005. "Sustainability indicators for the UK food supply chain." In: *Measuring Sustainability of the Food Supply Chain Seminar,* Cardiff University, October 27, 2005.

Zimmerman, J. *Sustainable Development through the Principles of Green Engineering.* Available from http://web.mit.edu/d-lab/assignment_files/ reen.pdf. Accessed September 19, 2008.

Index

Printed in the USA/Agawam, MA
May 7, 2013

575072.019